Optics: An Introduction

Optics: An Introduction

Summer Nolan

NY RESEARCH PRESS
New York

Published by NY Research Press
118-35 Queens Blvd., Suite 400,
Forest Hills, NY 11375, USA
www.nyresearchpress.com

Optics: An Introduction
Summer Nolan

International Standard Book Number: 978-1-63238-887-2 (Hardback)

Cataloging-in-Publication Data

Optics : an introduction / Summer Nolan.
p. cm.
Includes bibliographical references and index.
ISBN 978-1-63238-887-2
1. Optics. 2. Light. 3. Physics. I. Nolan, Summer.
QC355.3 .O68 2022
535--dc23

Table of Contents

Preface

It is with great pleasure that I present this book. It has been carefully written after numerous discussions with my peers and other practitioners of the field. I would like to take this opportunity to thank my family and friends who have been extremely supporting at every step in my life.

Optics studies the behavior and properties of light such as its interactions with matter as well as the construction of instruments that are used to detect it. It describes the behavior of visible, ultraviolet and infrared light. The two major areas of study within classical optics are physical optics and geometric optics. Physical optics considers light to propagate as a wave and geometric optics believes that light travels in straight lines as rays. Modern optics studies electromagnetic and quantum properties of light. Optics is used in various disciplines such as engineering, astronomy, photography and medicine. This book presents the complex subject of optics in the most comprehensible and easy to understand language. Some of the diverse topics covered herein address the varied branches that fall under this category. Those with an interest in this field would find this book helpful.

The chapters below are organized to facilitate a comprehensive understanding of the subject:

Chapter - What is Optics?

Optics is the branch of physics that studies the properties and behavior of light and its interaction with matter. Quantum optics, crystal optics, thin-film optics, nonlinear optics etc. are some of the types that fall within it. This is an introductory chapter which will briefly introduce these types of optics.

Chapter - Geometrical Optics

Geometrical optics is a model of optics that describes light propagation in terms of rays. Laws of reflection, refraction, total internal reflection, etc. are studied under geometrical optics. This chapter has been carefully written to provide an easy understanding of geometrical optics.

Chapter - Wave Optics

Wave optics is the branch of optics which deals with the study of interference, diffraction and polarization. Huygen's principle, Doppler effect, wavefront and wave theory of light are some of its aspects. The topics elaborated in this chapter will help in gaining a better perspective about these aspects related to wave optics.

Chapter - Light as an Electromagnetic Wave

The electromagnetic wave nature of light includes the Maxwell's equations, Malus' law, the electromagnetic spectrum, polarization of light, etc. This chapter closely examines these key concepts associated with the electromagnetic wave nature of light to provide an extensive understanding of the subject.

Chapter - Particle Nature of the Light

The photoelectric effect refers to the emission of electrons when a ray of light strikes a material. Compton effect is the scattering of photons by an electron which results in the decrease of energy of the photons. This chapter delves into the subject related to the particle nature of light for an in-depth understanding.

Chapter - The Optical Fiber

Optical fiber is a transparent fiber made from silica and plastic that is widely used in fiber-optic communications. It can be categorized into single-mode optical fiber and multi-mode optical fiber. This chapter has been carefully written to provide an easy understanding of these types of optical fibers.

Chapter - Understanding Lasers

The device that produces light through the process of optical amplification is known as a laser. Some of the types of lasers are solid-state lasers, liquid lasers, gas lasers and semiconductor lasers. This chapter discusses the characteristics and working of these types of lasers.

Summer Nolan

1

What is Optics?

Optics is the branch of physics that studies the properties and behavior of light and its interaction with matter. Quantum optics, crystal optics, thin-film optics, nonlinear optics etc. are some of the types that fall within it. This is an introductory chapter which will briefly introduce these types of optics.

Optics is a branch of physics that deals with the determination of behaviour and the properties of light, along with its interactions with the matter and also with the instruments which are used to detect it.

Optics, in a simple manner, is used to describe the behaviour of visible light, infrared light, and the ultraviolet. Imaging is done with the help of a system called an image forming an optical system.

Light and its Optical Properties

Light is a form of energy which is in the form of an electromagnetic wave and is almost everywhere around us. The visible light has wavelengths measuring between 400–700 nanometres. Sun is the primary source of light by which plants utilize this to produce their energy.

In physics, the term light also refers to electromagnetic radiation of different kinds of wavelength, whether it is visible to the naked eye or not. Hence by this, the gamma rays, microwaves, X-rays and the radio waves are also types of light.

Reflection

Reflection is one of the primary properties of light. Reflection is nothing but what you see the images in the mirrors. Reflection is defined as the change in direction of light at an interface in-between two different media so that the wave-front returns into a medium from which it was originated. The typical examples for reflection of light include sound waves and water waves.

Speed of Light

The rate at which the light travels in free space is called Speed of light. For example, the light travels 30% slower in the water when compared to vacuum.

Refraction

The bending of light when it passes from one medium to another is called Refraction. This property of refraction is used in a number of devices like microscopes, magnifying lenses, corrective lenses, and so on. In this property, when the light is transmitted through a medium, polarization of electrons takes place which in-turn reduces the speed of light, thus changing the direction of light.

Total Internal Reflection

When a beam of light strikes the water, a part of the light is reflected, and some part of the light is refracted. This phenomenon is called as Total internal reflection.

Dispersion

It is a property of light, where the white light splits into its constituent colours. The phenomenon of dispersion can be observed in the form of a prism.

Applications of Optics

The properties of optics are applied in various fields of Physics:

- The refraction phenomenon is applied in the case of lenses (Convex and concave) for the purpose of forming an image of the object.
- Geometrical optics is used in studies of how the images form in an optical system.
- In medical applications, it is used in the optical diagnosis of the mysteries of the human body.
- It is used in the therapeutical and surgeries of the human tissues.

Nonlinear Optics

Nonlinear optics provides the basis of many key technologies used today for providing radiation sources in various wavelength ranges from x-rays to the terahertz (THz). The relative bandwidths of such sources range from $\Delta\nu/\nu \sim 10^{-15}$ to several octaves (i.e. $\Delta\nu/\nu \sim 1-10$). Nonlinear optical technologies are usually used in combination with laser technology, and the main aspect is to shift or extend the limited wavelength range directly accessible by the laser source.

Nonlinear optics is the study of phenomena that occur as a consequence of the modification of the optical properties of a material system by the presence of light. Nonlinear optical phenomena are "nonlinear" in the sense that they occur when the response of a material system to an applied optical field depends in a nonlinear manner on the strength of the optical field. Typically, only laser light is sufficiently intense to modify the optical properties of a material system. The beginning of the field of nonlinear optics is often taken to be the discovery of second-harmonic generation by Franken, shortly after the demonstration of the first working laser by Maiman in 1960.

There are many excellent books on nonlinear optics. A very good introduction is given by New. A more in-depth introduction is given by Boyd.

John Kerr (Glasgow, UK) demonstrated in 1875 that the refractive index of a number of solids and liquids is slightly changed by the application of a strong DC field U_{DC} : $\Delta n \propto U_{DC}^2$. This phenomenon, now known as the (DC) Kerr effect, was the first nonlinear optical effect to be observed.

Friedrich Pockels (University of Göttingen, Germany) studied in 1894 a related process known today as the Pockels or electro-optic effect, where the refractive index change is proportional to the field: $\Delta n \propto U_{DC}$. Whereas the Kerr effect is observable in liquids and amorphous solids, the Pockels effect occurs only in crystalline materials that lack a centre of symmetry.

Further substantial progress was made only much later, when sufficiently intense light sources became available. This happened with the invention and demonstration of the laser in 1960. The operation of many lasers relies also on nonlinear optical processes, for example on optical pumping or Kerr-lens mode locking.

Peter Franken and coworkers (University of Michigan, USA) carried out the first laser-driven nonlinear optical experiment in 1961. Their demonstration of second harmonic generation of light by a ruby laser pulse in a quartz crystal marks the origin of nonlinear optics as a new separate scientific subfield. The straightforward experimental arrangement of this demonstration is shown in figure.

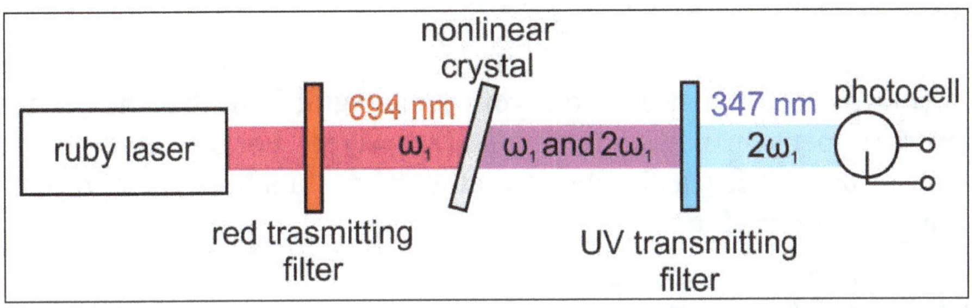

The scheme of the second-harmonic generation experiment of Franken and coworkers.

The Nobel Prize in Physics 1981 was divided, one half jointly to Nicolaas Bloembergen and Arthur Leonard Schawlow "for their contribution to the development of laser spectroscopy" and the other half to Kai M. Siegbahn "for his contribution to the development of high-resolution electron spectroscopy". Since its beginning, nonlinear optics keeps being an active field of research and technological development with an ever increasing number of applications.

Some Mathematical Tools

There are a few linear transformations often encountered in optics, which transform a function into another function (e.g. the Fourier transformation) or into a number (e.g. the Dirac delta). We review these basic operations, also in order to clarify notational conventions.

The Fourier transform of a function $f(t)$ is given by:

$$F_{\omega}\left[f(t)\right]=f(\omega):=\frac{1}{2\pi}\int_{-\infty}^{\infty}\mathrm{d}t\,f(t)e^{-i\omega t}$$

In solving differential equations, such as the wave equation, it is often helpful to use Fourier transformation. This explains the importance of this mathematical tool in optics. The inverse Fourier transform is:

$$F_{t}^{-1}\left[f(\omega)\right]=f(t)=\int_{-\infty}^{\infty}\mathrm{d}\omega\,f(\omega)e^{i\omega t}$$

It can be interpreted as the composition of the signal $f(t)$ from harmonic basis functions (Fourier components), with (complex) amplitudes $f(\omega)$. We note that a time-periodic function has discrete Fourier component spectrum.

The Dirac delta is defined as follows:

$$D_{t_0}\left[f(t)\right]:=f(t_0)$$

It is usually written in an integral form with the help of the Dirac delta 'function':

$$\int_{-\infty}^{\infty}\mathrm{d}t\,\delta(t-t_0)f(t):=D_{t_0}\left[f(t)\right]:=f(t_0)$$

It holds $\delta(t)=\delta(-t)$. It is also visible from the integral form that the delta function has the inverse dimension of its argument: $\left[\delta(t)\right]=1/\left[t\right]$. For the Dirac delta function, mathematically being in fact a distribution rather than a function, the following Fourier representation is also useful:

$$\delta(t)=\frac{1}{2\pi}\int_{-\infty}^{\infty}\mathrm{d}\omega e^{i\omega t}=\frac{1}{2\pi}\int_{-\infty}^{\infty}\mathrm{d}\omega e^{-i\omega t}$$

The Nonlinear Polarization

The constitutive relation $D(E) = \varepsilon_0 E + P(E)$ plays a key role in optics, especially in nonlinear optics. It describes the response of a medium to the applied external electric field. Nonlinear optics deals with cases where this response is a nonlinear function of the external field.

Nonlinear Response: Instantaneous Case

For higher field strengths, the material response to the applied field may become nonlinear, which can be described by a Taylor-series expansion:

$$\begin{aligned} P(E) &= \varepsilon_0 \left(\chi^{(1)} E + \chi^{(2)} EE + \chi^{(3)} EEE + \cdots \right) \\ &= P^{(1)} + P^{(2)} + P^{(3)} + \cdots \\ &= P^{(1)} + P^{NL} \end{aligned}$$

Here $\chi^{(n)}$ is the nth-order nonlinear susceptibility, a tensor of rank $n+1$ with 3^n components. The above equation can be written in matrix form as follows:

$$P_i(E) = \varepsilon_0 \left(\sum_j \chi_{ij}^{(1)} E_j + \sum_{j,k} \chi_{ijk}^{2} E_j E_k + \sum_{j,k,l} \chi_{ijkl}^{(3)} E_j E_k E_l + \cdots \right) (i, j, k, l = x, y, z)$$

Similarly to the case of linear response, the above equations correspond to the case of instantaneous response to a time-dependent electric field, meaning also that $\chi^{(1)}, \chi^{(2)}, \chi^{(3)}, \ldots$ are constants.

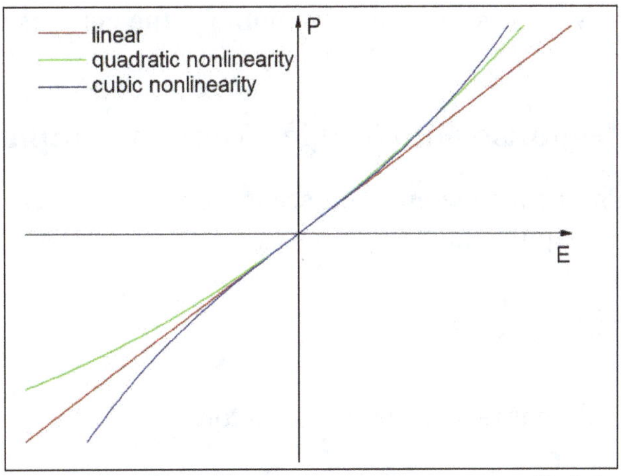

Linear, quadratic, and cubic polarization response to an applied electric field.

In figure illustrates the various contributions to the polarization in response to an applied electric field. In case of linear response the polarization is proportional to the electric field. In case of quadratic (2nd-order) and cubic (3rd-order) nonlinearity the response deviates from simple proportionality. Figure illustrates the response of materials to an applied harmonic external electric field. In case of linear response, the polarization oscillates with the same frequency as the driving field. In the nonlinear case, new frequency components appear. Please note the distinctly different form of the responses for centrosymmetric and non-centrosymmetric crystal structures.

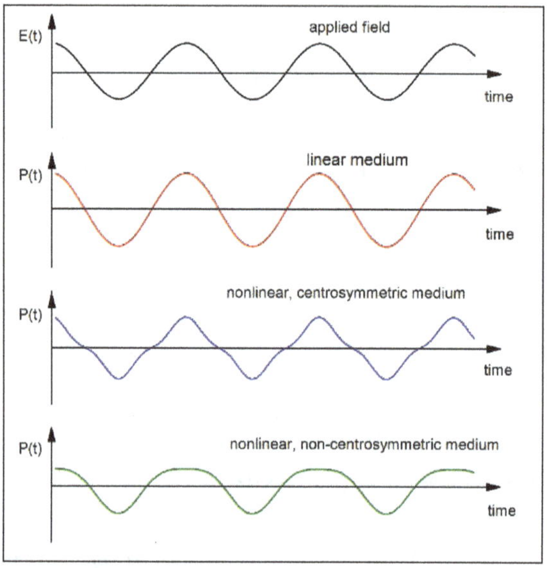

Illustration of various types of material responses to an applied harmonic field.

During propagation of a light beam in a medium with nonlinear response new frequency components can be generated which were not contained in the input. To give a first impression of the many possibilities which can arise from such an interaction we give below a few examples in a simple qualitative treatment. To simplify the treatment as much as possible we will use scalar notation both for the fields as well as for the nonlinear susceptibility.

Second-Order Response with Single-Frequency Input

Let us consider a nonlinear medium with second-order nonlinear susceptibility $\chi^{(2)}$ and a monochromatic input field:

$$E(t) = \hat{E}\cos(\omega t) = \frac{1}{2}\hat{E}\,e^{i\omega t} + \text{c.c.}$$

The second-order nonlinear response is given as follows:

$$P^{(2)}(t) = \varepsilon_0 \chi^{(2)} EE = \frac{1}{2}\varepsilon_0 \chi^{(2)} \hat{E}\hat{E}^* + \frac{1}{4}\varepsilon_0 \chi^{(2)} \hat{E}^2 e^{i2\omega t} + \text{c.c.}$$

The first term describes a constant (DC) field, the second one a field which oscillates at twice the input frequency. We speak of optical rectification (OR) and second-harmonic generation (SHG), respectively. The relation between the driving field and the material response is illustrated in figure. The scheme of the SHG process driven by a laser beam of fundamental frequency ω is shown in figure.

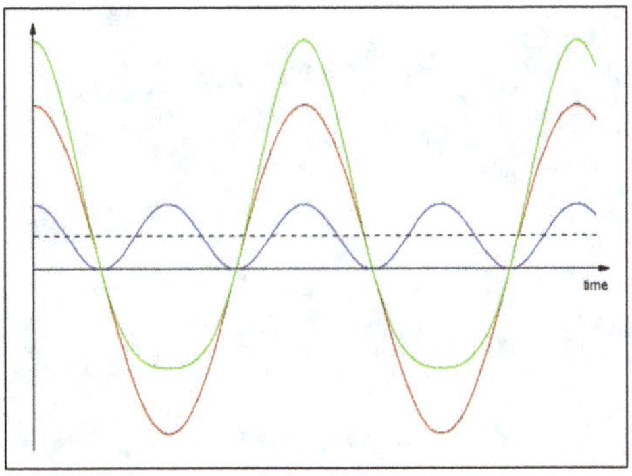

Response (green line) of a medium with second-order nonlinearity to a harmonic driving field (red line). The response contains both second-harmonic (blue line) and DC (black dashed line) components.

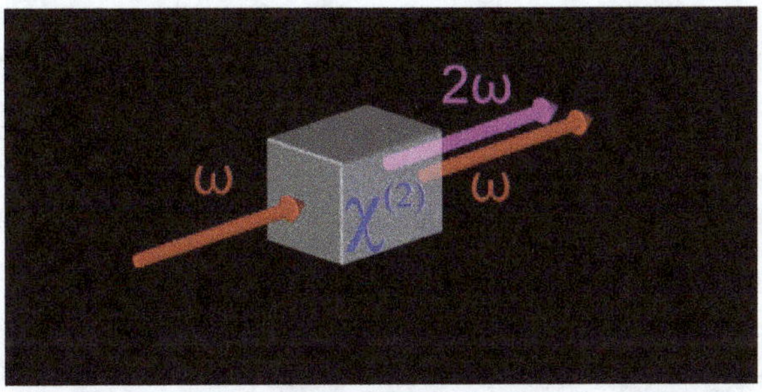

Scheme of SHG.

Second-Order Response with Two-Frequency Input

Let us consider a nonlinear medium with second-order nonlinear susceptibility $\chi^{(2)}$ and an input field which contains two distinct frequency components:

$$E(t) = \frac{1}{2}\hat{E}_1 e^{i\omega_1 t} + \frac{1}{2}\hat{E}_2 e^{i\omega_2 t} + \text{c.c.}$$

The second-order nonlinear response is given as follows:

$$P^{(2)}(t) = \varepsilon_0 \chi^{(2)} EE = \frac{1}{4}\varepsilon_0 \chi^{(2)} \left[\hat{E}_1^2 e^{i2\omega_1 t} + \hat{E}_2^2 e^{i2\omega_2 t} + 2\hat{E}_1 \hat{E}_2 e^{i(\omega_1+\omega_2)t} + 2\hat{E}_1 \hat{E}_2^* e^{i(\omega_1+\omega_2)t} + 2\hat{E}_1 \hat{E}_1^* \right] + \text{c.c}$$

The first two terms describe SHG of both frequency components, the third term describes sum-frequency generation (SFG), the fourth term is difference-frequency generation (DFG), while the last term is OR. We note that in practice a measurable output field will be generated usually just at one of the many possible new frequency components. The reason is that a fixed phase relationship (phase matching) should be maintained during the propagation which cannot be fulfilled simultaneously for all frequency combinations. The schemes of SFG and DFG are shown in figures.

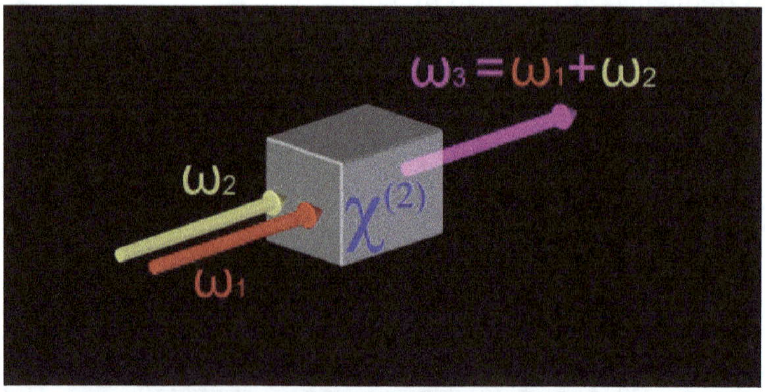

Scheme of SFG.

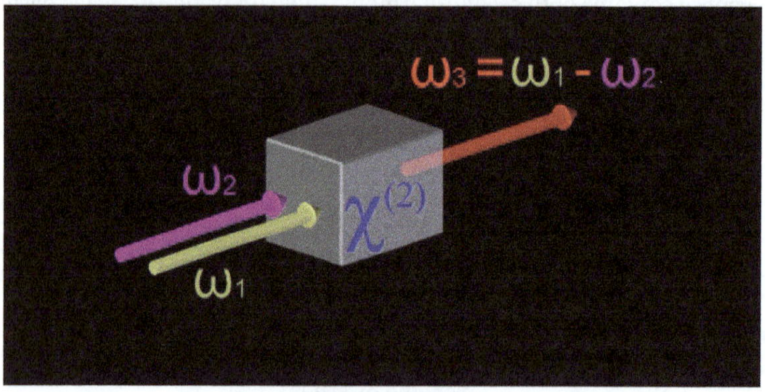

Scheme of DFG.

Third-order Response with Single-frequency Input

Let us consider a nonlinear medium with third-order nonlinear susceptibility $\chi^{(3)}$ and a monochromatic input field:

$$E(t) = \hat{E}\cos(\omega t) = \frac{1}{2}\hat{E}\,e^{i\omega t} + \text{c.c.}$$

The third-order nonlinear response is given as follows:

$$P^{(3)}(t) = \varepsilon_0 \chi^{(3)} EEE = \frac{1}{8}\varepsilon_0 \chi^{(3)} \hat{E}^3 e^{i3\omega t} + \frac{3}{8}\varepsilon_0 \chi^{(3)} \hat{E}^2 \hat{E}^* e^{i\omega t} + \text{c.c.}$$

The first term describes a field which oscillates at the frequency 3ω and we speak of third-harmonic generation (THG). The scheme of THG is shown in figure. The second term is a nonlinear contribution to the field component at the input frequency ω and is called the AC or optical Kerr effect. This results in an intensity-dependent change of the refractive index of the medium.

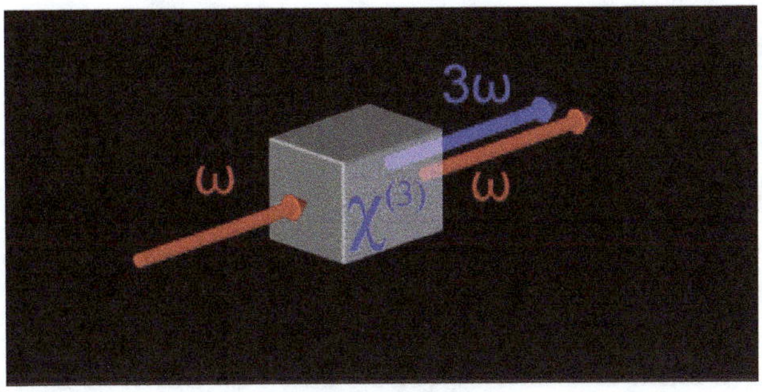

Scheme *of THG.*

The refractive index change can be given by considering both linear and nonlinear contributions to the polarization at frequency ω:

$$\begin{aligned}P(t) &= P^{(1)}(t)+P_{\omega}^{(3)}(t)\\ &= \varepsilon_0\chi^{(1)}\frac{1}{2}\hat{E}\,e^{i2\omega t}+\frac{3}{8}\varepsilon_0\chi^{(3)}\hat{E}^2\hat{E}^*e^{i\omega t}+\text{c.c.}\\ &= \varepsilon_0\left(\chi^{(1)}+\frac{3}{4}\chi^{(3)}\hat{E}\hat{E}^*\right)\frac{1}{2}\hat{E}e^{i\omega t}+\text{c.c.}\\ &= \varepsilon_0\chi\frac{1}{2}\hat{E}e^{i\omega t}+\text{c.c.}\end{aligned}$$

with $n=\sqrt{1+\chi}=\sqrt{1-\chi^{(1)}-(3/4)\chi^{(3)}\hat{E}\hat{E}^*}$. The induced change in refractive index can be estimated from $\Delta n=\sqrt{1+\chi}-\sqrt{1+\chi^{(1)}}$ and $I=c\varepsilon_0\,r,\left|\hat{E}\right|^2/2$ as

$$\Delta n\approx\frac{3\chi^{(3)}}{4c\varepsilon_0 n^2}I$$

Nonlinear Response: Medium with Memory

For real materials with non-instantaneous and nonlinear response, a similar procedure can be applied to that given above for the linear case. For the second-order nonlinear term the time-domain form of the constitutive relation can be written as follows:

$$\begin{aligned}\mathrm{P}^{(2)}(t) &= \varepsilon_0\int_{-\infty}^{\infty}d\tau\int_{-\infty}^{\infty}d\tau'\rho^{(2)}(\tau,\tau')\mathrm{E}(t-\tau)\mathrm{E}(t-\tau')\\ &= \varepsilon_0\int_{-\infty}^{\infty}d\omega\int_{-\infty}^{\infty}d\omega'\chi^{(2)}(\omega,\omega')\mathrm{E}(\omega)\mathrm{E}(\omega')e^{i}(\omega+\omega')t\end{aligned}$$

The second-order nonlinear response function $\rho^{(2)}(\tau,\tau')$ is a third-rank tensor quantity with dimension $\left[\rho^{(2)}\right]=\mathrm{Hz}^2.m/V$. To preserve causality $\rho^{(2)}(\tau,\tau')=0$ must hold for $\tau, \tau' < 0$. The second-order nonlinear susceptibility, also a tensor of rank three, is defined by the two-dimensional Fourier transform of $\rho^{(2)}$:

$$\chi^{(2)}_{i,j,k}(\omega,\omega') = \frac{1}{(2\pi)^2} \int_{-\infty}^{\infty} d\tau \int_{-\infty}^{\infty} d\tau' (2\pi)^2 \rho^{(2)}_{i,j,k}(\tau,\tau') e^{-i(\omega\tau+\omega'\tau')} \quad (i,j,k = x,y,z)$$

The dimension of $\chi^{(2)}(\omega,\omega')$ is m/V. The general frequency-domain relation corresponding to the above time-domain equation can be obtained by Fourier transformation. We will consider some important special cases later.

Coupled Wave Equations

In many cases encountered in nonlinear optics it is possible to decompose the electromagnetic field into a discrete set of (quasi-monochromatic) frequency components:

$$\mathrm{E}(r,t) = \frac{1}{2} \sum_n \hat{\mathrm{E}}_n(r,t) e^{i\omega nt} + \text{c.c.}$$

Here $\hat{\mathrm{E}}_n$ is the complex amplitude of the frequency component at ω_n. It is slowly varying in time (unless we deal with extremely short pulses containing only very few oscillations of the electromagnetic field), but in space it varies on the scale of the optical wavelength. Note that $\hat{\mathrm{E}}_n(r,t)$ is not the Fourier amplitude of the field; accordingly, its dimension is $\hat{\mathrm{E}}_n = \mathrm{V/m}$. For this reason we have used the words 'frequency component' above rather than 'spectral component'. We reserve the latter wording for the components of the field's Fourier transform. The summation in the above expression shall include all frequency components present in the input field as well as those newly generated inside the medium owing to the nonlinear response.

A typical example for a discrete set of frequency components is the propagation of one or more monochromatic laser beams in a nonlinear medium, where additional frequency components may also be generated. Here, the summation should include both the incoming and the possibly generated new frequency components. In this case $\hat{\mathrm{E}}_n$ is constant in time. Another example is the nonlinear propagation of one or more pulsed laser beams in a nonlinear medium. In this case the summation should include the frequency components of the incoming pulse(s) and the new frequency components eventually generated in the nonlinear medium.

Monochromatic Frequency Components

In case of monochromatic frequency components $\hat{\mathrm{E}}_n(r,t) = \hat{\mathrm{E}}_n(r)$ and the field can be written as follows:

$$\mathrm{E}(r,t) = \frac{1}{2} \sum_n \hat{\mathrm{E}}_n(\mathrm{r}) \mathrm{e}^{i\omega nt} + \text{c.c.}$$

We note that the Fourier transform of $\mathrm{E}(r,t)$ is given by:

$$\mathrm{E}(r,\omega)=\frac{1}{2}\sum_{n}\left[\hat{\mathrm{E}}_n(\mathrm{r})\delta(\omega-\omega_n)+\hat{\mathrm{E}}_n^{*}(\mathrm{r})\delta(\omega+\omega_n)\right]$$

It can be shown easily that in this case the linear response is also composed of monochromatic frequency components:

$$\mathrm{P}^{(1)}(r,t)=\frac{1}{2}\sum_{n}\left[\hat{\mathrm{P}}_n^{(1)}(r)e^{i\omega_n t}+\mathrm{c.c.}\right]$$

where $\hat{\mathrm{P}}_n^{(1)}=\varepsilon_0\chi^{(1)}(\omega_n)\hat{\mathrm{E}}_n$ is the complex amplitude of the linear polarization component with frequency ω_n. Similarly to the electric field, $\hat{\mathrm{P}}_n^{(1)}(r,t)$ has the same dimension as $\hat{\mathrm{P}}^{(1)}(r,t)$, i.e. $\mathrm{C/m^2}$, and therefore it is not a spectral amplitude. Similarly, there can be nonlinear contributions to the polarization at each frequency:

$$\mathrm{P}^{\mathrm{NL}}(r,t)=\frac{1}{2}\sum_{n}\left[\hat{\mathrm{P}}_n^{\mathrm{NL}}(r)e^{i\omega_n t}+\mathrm{c.c.}\right]$$

One can substitute the above formulas into the wave equation, which then separates into individual equations for each complex amplitude:

$$\nabla^2\hat{\mathrm{E}}_n(r)+\frac{\omega_n^2}{c^2}\varepsilon(\omega_n)\hat{\mathrm{E}}_n(r)=-\frac{\omega_n^2}{\varepsilon_0 c^2}\hat{\mathrm{P}}_n^{\mathrm{NL}}(r)$$

Thus, for a field with N discrete frequency components one has N separate amplitude equations. For linear media, the amplitude equations are homogenous, and completely decoupled from each other. As a consequence, no new frequency components are generated during the propagation in the medium, and there is no interaction between the components. The situation can be drastically changed in case of nonlinear response. The amplitude equations are no more homogeneous; it is the amplitude of the nonlinear part of the polarization that shows up on the right-hand side. Depending on the type of the nonlinear interaction, it can act as the source of new frequency components which were not present in the field entering the medium. The individual amplitude equations may become coupled to each other, and field components with different frequencies will interact during propagation. The set of amplitude equations describing the interacting field components are often called coupled wave equations. They play a central role in nonlinear optics. The nonlinear interaction of optical waves leads to a very rich world of new phenomena.

Monochromatic Spectral Components with Slowly-varying Amplitudes

In optics one often deals with radiation fields in the form of light beams. Except the cases of very strong focusing or defocusing, these beams change their field distributions or sizes on length scales much larger than the optical wavelength. This also means that their wavefronts are, to a good approximation, planes. Thus, in many situations it is adequate to describe the propagating light beams by waves with amplitudes which vary on length scales much larger than the optical wavelength. For a quasi-monochromatic frequency component with slowly-varying amplitude propagating in the direction of the wave vector k one can write:

$$\mathrm{E}(r,t) = \frac{1}{2}\tilde{\mathrm{E}}(r,t)e^{i(\omega t - \mathrm{kr})} + \mathrm{c.c.}$$

The angular wavenumber is $|\mathrm{k}| = k = 2\pi / \lambda = \omega . n(\omega)/c$ with the refractive index $n(\omega) = \sqrt{\varepsilon(\omega)}$. The complex amplitude $\tilde{\mathrm{E}}(r,t)$ is slowly varying both in time as well as in space. It holds $\tilde{\mathrm{E}}(r,t) = \tilde{\mathrm{E}}(r,t)e^{-ikr}$. We note that for a monochromatic plane wave with constant amplitude, the field can be written as $\mathrm{E}(r,t) = \mathrm{E}_0 \cos(\omega t - \mathrm{kr} + \phi_0)$, and $\tilde{\mathrm{E}}(r,t) = \mathrm{E}_0 e^{i\phi_0}$ holds.

A field composed of several monochromatic components with slowly-varying amplitudes can be written as follows:

$$\hat{\mathrm{E}}(r,t) = \frac{1}{2}\sum_n \tilde{\mathrm{E}}_n(\mathrm{r})e^{i(\omega_n t - \mathrm{k}_n \tau)} + \mathrm{c.c.}$$

$$\mathrm{P}^{\mathrm{NL}}(r,t) = \frac{1}{2}\sum_n \tilde{\mathrm{P}}_n^{\mathrm{NL}}(r)e^{i(\omega_n t - \mathrm{k}_n \tau)} + \mathrm{c.c}$$

For waves propagating in the z-direction (which means no loss of generality unless one deals with non-collinear geometry) we can substitute,

$$\hat{\mathrm{E}}_n(r) = \tilde{\mathrm{E}}_n(r)e^{-ik_n z}$$

$$\hat{\mathrm{P}}_n^{\mathrm{NL}}(r) = \hat{\mathrm{P}}_n^{\mathrm{NL}}(r)e^{-ik_n z}$$

into the coupled wave equations and obtain the corresponding coupled equations for the slowly varying amplitudes:

$$2ik_n \frac{\partial}{\partial z}\tilde{\mathrm{E}}_n(r) - \frac{\partial^2}{\partial z^2}\tilde{\mathrm{E}}_n(r) - \nabla\frac{2}{T}\tilde{\mathrm{E}}_n(r) = \frac{\omega_n^2}{\varepsilon_0 c^2}\tilde{\mathrm{P}}_n^{\mathrm{NL}}(r)$$

For slowly varying amplitudes the second-order -derivative can be neglected, since it is much smaller than the first-order term. The term with $\nabla\frac{2}{T} = \partial_x^2 + \partial_y^2$ describes diffraction of the beam. This term vanishes for plane waves since their amplitudes depend only on the z-coordinate: $\tilde{E}_n(r) = \tilde{E}_n(z)$. For such case, with $n_n = n_n(w_n)$, one obtains the following form of the coupled wave equations:

$$\frac{\partial^2}{\partial z^2}\tilde{E}_n(z) = i\frac{\omega_n}{2c\varepsilon_0 n_n}\tilde{P}_n^{NL}(z)$$

This is the simplest form of the coupled wave equations in nonlinear optics. It shows that $\tilde{E}_n$ will grow at a maximal rate if the polarization wave has a phase lead of $\pi/2$ with respect to the field (i.e. the arguments of $\tilde{P}_n^{NL}$ and $\tilde{E}_n$ are the same). If the sign of the phase difference is reversed (a phase lag of $\pi/2$), the field will decay at a maximal rate. If the polarization and field are in phase, the amplitude of the field will be unchanged, but its phase will change under propagation. In this case, it is the angular wave number and hence the refractive index that are modified.

We shall now investigate in detail the form of the second-order nonlinear polarization. We consider the field to be composed of monochromatic frequency components with slowly varying amplitudes:

$$E(z,t) = \frac{1}{2}\sum_n \hat{E}_n(z)e^{i\omega_n t} + \text{c.c} = \frac{1}{2}\sum_n \tilde{E}_n(z)e^{i(\omega_n t - k_n z)} + \text{c.c.}$$

Here we have assumed that all components are plane waves propagating in the z-direction. The second-order nonlinear response can be given as follows:

$$P^{(2)}(z,t) = \varepsilon_0 \int_{-\infty}^{\infty} d\omega \int_{-\infty}^{\infty} d\omega' \chi^{(2)}(\omega,\omega')E(z,\omega)E(z,\omega')e^{i(\omega+\omega')t}$$

The Fourier amplitudes:

$$E(z,\omega) = \frac{1}{2}\sum_n \left[\hat{E}_n(z)\delta(\omega-\omega_n)\hat{E}_n^*(z)\delta(\omega+\omega_n)\right]$$

can be substituted into the above expression to give:

$$P^{(2)}(z,t) = \frac{\varepsilon_0}{4}\sum_{n,m}\left[\chi^{(2)}(\omega_n,\omega_m)\hat{E}_n(z)\hat{E}_m(z)e^{i(\omega_n+\omega_m)t} + \chi^{(2)}(\omega_n,-\omega_m)\hat{E}_n(z)\hat{E}_m(z)e^{i(\omega_n-\omega_m)t}\right]$$

Here we have made use of the symmetry property of the nonlinear susceptibility components $\chi^{(2)}(-\omega_n,-\omega_m) = \chi^{(2)*}(\omega_n,\omega_m)$ owing to the reality of the fields. In words: changing the sign of both frequency arguments is equivalent to taking the complex conjugate.

The nonlinear polarization at each particular frequency component can be obtained by gathering all terms from the above equation containing the respective frequency. In practical situations it is usually only a rather limited number of frequency components which play a role. A few important examples will be studied below.

Second-harmonic Generation (SHG)

In case of second-harmonic generation (SHG) we study the interaction of a driving field of frequency ω_1 and its second-harmonic field at the doubled (second harmonic, SH) frequency $\omega_2 = 2\omega_1$. Thus the field inside the nonlinear medium consists of only two monochromatic frequency components:

$$\mathrm{E}(z,t) = \frac{1}{2}\sum_{n=1}^{2}\hat{\mathrm{E}}_n(z)e^{i\omega_n t} + \mathrm{c.c.} = \frac{1}{2}\sum_{n=1}^{2}\tilde{\mathrm{E}}_n(z)e^{i(\omega_n t - k_n z)} + \mathrm{c.c.}$$

There is only one term in the expression for second-order nonlinear polarization above, which describes a field at ω_2; it is the one with indices $n - m = 1$. There are two terms describing the nonlinear polarization at the fundamental frequency ω_1, those with $(n,m) = (1,2)(2,1)$. By substituting the expressions for these nonlinear polarizations into the coupled wave equations, the following form is obtained,

$$\frac{\partial}{\partial z}\tilde{\mathrm{E}}_1(z) = -\mathrm{i}\frac{\omega_1}{2cn_1}\chi^{(2)}(\omega_2,-\omega_1)\tilde{\mathrm{E}}_2(z)\tilde{\mathrm{E}}_1^*(z)e^{-i\Delta kz}$$

$$\frac{\partial}{\partial z}\tilde{\mathrm{E}}_2(z) = -\mathrm{i}\frac{\omega_2}{4cn_2}\chi^{(2)}(\omega_1,\omega_1)\tilde{\mathrm{E}}_1(z)\tilde{\mathrm{E}}_1(z)e^{i\Delta kz}$$

Here, $\Delta k = k_2 - 2k_1 = (n_2 - n_1)2\omega_1 / c$ is the phase-mismatch parameter. Please note the factor of 2 difference in the coefficients on the right-hand side, which can be traced back to the number of terms contributing to the nonlinear polarizations.

In case of a lossless medium the full permutation of the nonlinear susceptibility applies, and one can use the same $\chi^{(2)} =: \chi^{\mathrm{SHG}}$ in both equations. For such a case the above equations imply,

$$\frac{dI_1}{dz} = \frac{dI_2}{dz}$$

which is expressing energy conservation for the SHG process. It states that the total intensity $I = I_1 + I_2$ is constant. Obviously, this relation should be modified for a medium with loss.

The photon current density, i.e. the number of photons transported through unit surface within unit time interval, can be defined as $\Phi_i = I_i / \hbar\omega_i \, (i=1,2)$. With this energy conservation can be formulated as:

$$\frac{d\Phi_1}{dz} = 2\frac{d\Phi_2}{dz}$$

This is called the Manley–Rowe relation for SHG, stating that the rate at which photons at frequency ω_1 are annihilated is twice the rate at which photons at frequency ω_2 are created. Please remember that photons are energy quanta of the radiation field. We also note that quantum mechanics is implicitly involved in the derivation of these equations through the assumption of full permutation symmetry.

Non-depleted Pump Approximation

In many practical cases there is a strong fundamental beam at the input side of the nonlinear medium which propagates nearly unattenuated through the medium while a weaker SH field is being built up. In this case one often considers $\tilde{E}_1(z)$ to be constant (independent of *z*) and the first of the two coupled wave equations can be omitted, and the second one easily solved. The solution for $\tilde{E}_2(z=0)=0$ is as follows:

$$\tilde{E}_2(z) = -i\frac{\omega_2}{4cn_2}\chi\,\text{SHG}\,\tilde{E}_1\tilde{E}_1\frac{e^{i\Delta kz}-1}{i\Delta k} = -i\frac{\omega_2}{4cn_2}\chi\,\text{SHG}\,\tilde{E}_1\tilde{E}_1\,z\,\text{sinc}\left(\frac{\Delta kz}{2}\right)e^{i\Delta kz/2}$$

where $\text{sinc}(x) = \sin(x)/x$.

In the simplest cases one has well-defined electric field polarization directions for both fundamental and SH beams. This allows us to simplify the notation and use scalar quantities for the electric fields and the nonlinear susceptibility, meaning the single non-zero Cartesian vector component and an effective nonlinear susceptibility. More general cases, highly relevant for many practical situations, will be considered later. Under these circumstances, by using the relation $I_{1,2}(z) = c\varepsilon_0 n_{1,2}\left|\tilde{E}_{1,2}(z)\right|^2/2$ for the fundamental and the SH fields, the SH intensity can be written as:

$$I_2(z) = \frac{1}{2}c\varepsilon_0 n_2\left|\tilde{E}_2(z)\right|^2 = \frac{\left(\omega_2\chi^{\text{SHG}}\right)^2}{8\varepsilon_0 c^3 n_2 n_1^2} I_1^2 z^2 \,\text{sin c}^2\left(\frac{\Delta kz}{2}\right)$$

The SH intensity is proportional to the square of both the nonlinear susceptibility, and the fundamental intensity. The fundamental-to-SH conversion efficiency is given by $I_2(z)/I_1$, and is also proportional to I_1.

For a fixed propagation distance the SH intensity depends on the phase-mismatch parameter Δk, as shown in figure. Since the argument of the sinc function is proportional to the propagation (interaction) distance z, the Δk-dependence is more pronounced for large interaction distances. Please note that the Δk-dependence also implies frequency dependence through material dispersion, which is typically much more sensitive than the ω_2^2-dependence of the pre-factor. On the other hand, the peak intensity scales with the square of the interaction distance. Therefore, large bandwidth and large conversion efficiency usually are conflicting goals. The interactive version of figure allows to vary the propagation distance while showing the SH intensity as function of Δk.

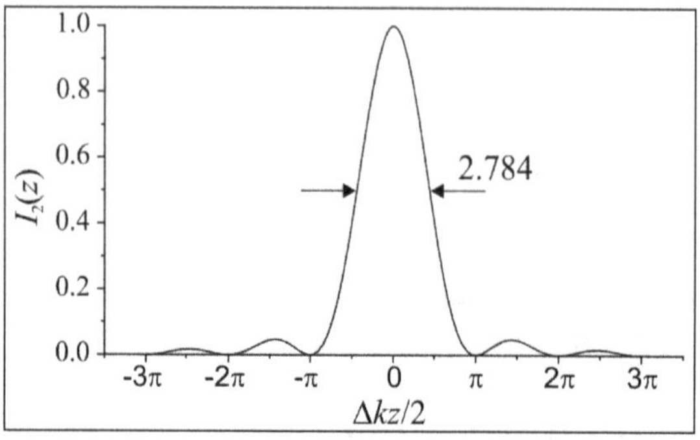

Dependence of the second-harmonic intensity on the phase-mismatch parameter.

From the above expression it is also clear that the maximum growth rate of the SH field can be achieved for $\Delta k = 0$, i.e. for perfect phase matching. In this case, the SH intensity increases quadratically with propagation distance *z*. Of course, at some point, the nondepleted pump approximation breaks down and pump attenuation will start to play a role.

In case of phase mismatch ($\Delta k \neq 0$), the dependence of the SH intensity on *z* becomes oscillatory:

$$I_2(z) = \frac{\left(\omega_2 \chi^{\mathrm{SHG}}\right)^2}{2\varepsilon_0 c^3 n_2 n_1^2} I_1^2 \frac{1}{\Delta k^2} \sin^2\left(\frac{\Delta kz}{2}\right) = I_2^{\max} \sin^2\left(\frac{\Delta kz}{2}\right)$$

The intensity reaches this maximum for the first time at the coherence length $z = L_{\mathrm{coh}}$:

$$L_{\mathrm{coh}} = \frac{\pi}{|\Delta k|}$$

The interactive figure shows the SH intensity as function of the propagation distance with Δk as the variable parameter.

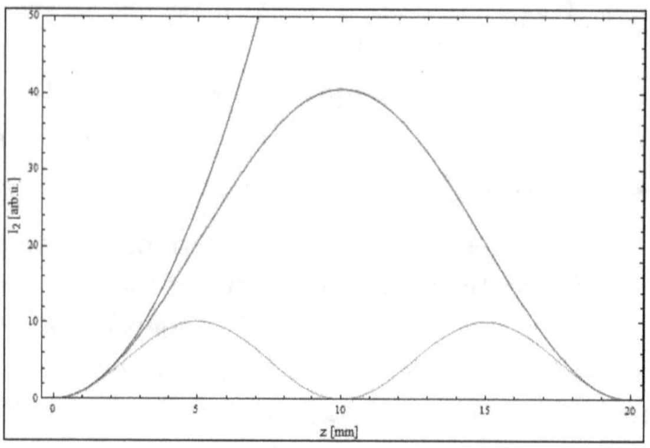

SH intensity as function of the propagation distance for phase-matched (blue line) and non-phase-matched (red and grey lines) conditions.

SHG in Practice

In practice, different experimental schemes can be used for SHG. For many applications it is essential to achieve high conversion efficiency. In efficient schemes, with continuous-wave (CW) or long-pulse (ns) lasers, experimental values can typically be as high as 30 to 50%, or even up to 80%. Clearly, such high efficiencies are well beyond the validity of the non-depleted pump approximation. Basic designs of SHG devices are as follows: (i) single-pass configuration, (ii) double-pass configuration, (iii) external-cavity configuration, and (iv) intracavity configuration.

A double-pass configuration can be seen in. By changing the mirror position on the right-hand side, the relative phase between the fundamental and the SH generated from the first pass through the nonlinear crystal can be varied. This is possible owing to the refractive index dispersion of air. By optimizing this relative phase the SH field originating from the second pass adds constructively to that from the first pass. This enables up to two times higher SH amplitude and, consequently, four times higher SH intensity at the output, as compared to single-pass SHG with the same crystal length.

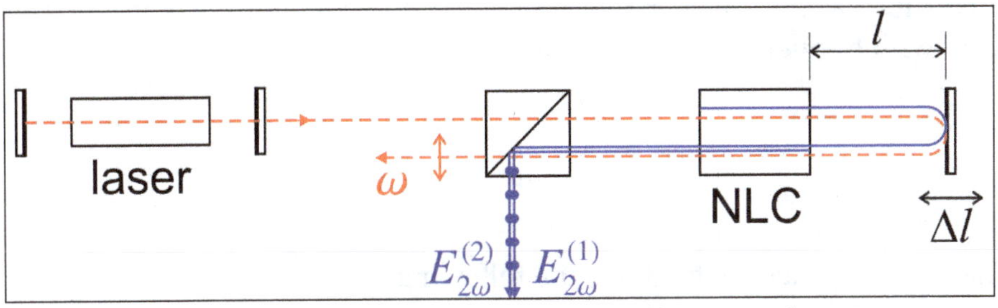

Scheme of double-pass SHG. NLC: nonlinear crystal.

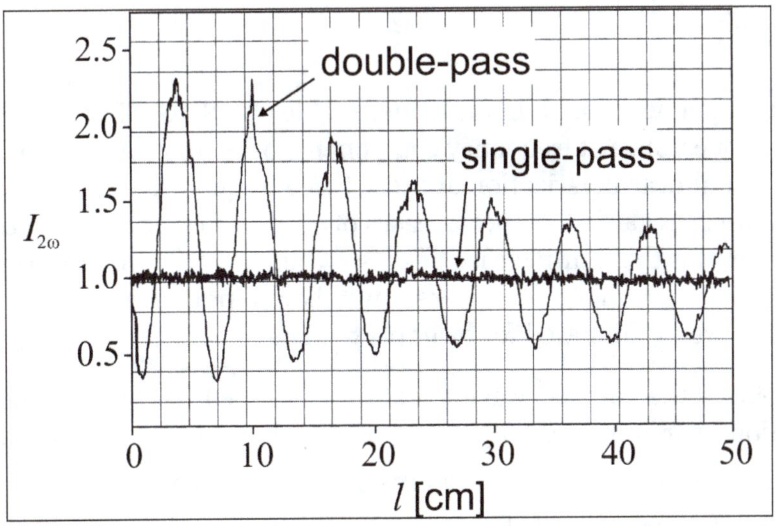

Experimental results for double-pass SHG in comparison with the single-pass geometry.

Two possible external-cavity and the intracavity configurations are shown in figure.

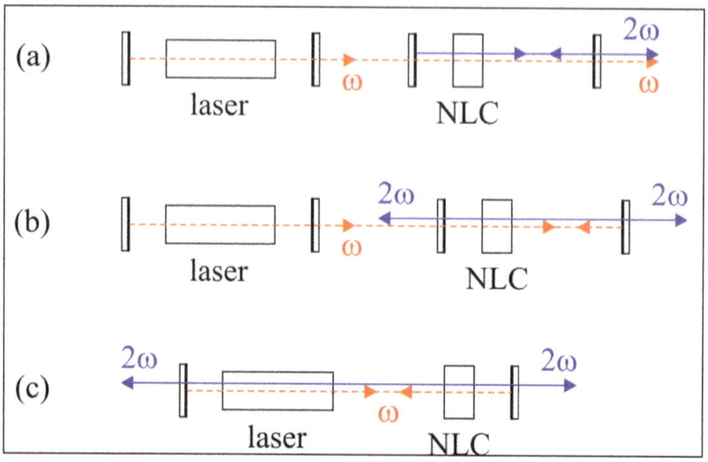

Various SHG configurations involving optical resonators. (a) External cavity resonating the SH. (b) External cavity resonating the fundamental. (c) Intracavity SHG. NLC: nonlinear crystal.

Three-wave Mixing (TWM)

Now we consider a more general case of second-order nonlinear interaction where three waves with frequencies ω_1, ω_2, and ω_3 are involved,

and,

$$\omega_3 = \omega_1 + \omega_2$$

holds. The incoming electric field has the following form:

$$\mathrm{E}(z,t) = \frac{1}{2}\sum_{n=1}^{3} \tilde{\mathrm{E}}_n(z) e^{i\omega_n t} + \mathrm{c.c.} = \frac{1}{2}\sum_{n=1}^{3} \tilde{\mathrm{E}}_n(z) e^{i(\omega_n t - k_n z)} + \mathrm{c.c.}$$

While such an input field could, in principle, enable OR, SHG, SFG, and DFG, thereby leading to the appearance of frequency components other than ω_1, ω_2, or ω_3, we consider only the nonlinear polarization at these three frequencies. It shall be noted that OR and SHG can be regarded as special cases of such a more general three-wave interaction process. OR is the limiting case of $\omega_3 \approx \omega_2$ and $\omega_1 \approx 0$, while SHG corresponds to $\omega_1 = \omega_2$. However, owing to the different number of input beams, some distinct difference occurs here as compared for example to SHG.

In order to obtain the coupled wave equations, we need to consider the second-order polarization at frequencies ω_1, ω_2, and ω_3. This can be obtained by substituting the above three-frequency input field into the expression of $\mathrm{P}^{(2)}(z,t)$. There are two terms contributing to the field at ω_1: those with $(n,m) = (3,2)(2,3)$, two terms to ω_2: $(n,m) = (3,1)(1,3)$, and two terms to ω_3: $(n,m) = (1,2)(2,1)$. By taking into account these contributions one obtains the following form of the coupled wave equations for three-wave mixing:

$$\frac{\partial}{\partial z}\tilde{E}_1(z) = -i\frac{\omega_1}{2cn_1}\chi^{(2)}(\omega_3,-\omega_2)\tilde{E}_3(z)\tilde{E}_2^*(z)e^{-i\Delta kz}$$

$$\frac{\partial}{\partial z}\tilde{E}_2(z) = -i\frac{\omega_2}{2cn_2}\chi^{(2)}(\omega_3,\omega_1)\tilde{E}_3(z)\tilde{E}_1^*(z)e^{-i\Delta kz}$$

$$\frac{\partial}{\partial z}\tilde{E}_3(z) = -i\frac{\omega_3}{2cn_3}\chi^{(2)}(\omega_1,\omega_2)\tilde{E}_1(z)\tilde{E}_2(z)e^{i\Delta kz}$$

Similarly to what was said for SHG, for lossless media full permutation symmetry holds and the same $\chi^{(2)} =: \chi^{\mathrm{TWM}}$ can be used in all of the three equations above. According to,

$$\frac{dI_1}{dz} = \frac{dI_2}{dz} = -\frac{dI_3}{dz}$$

the total intensity is conserved. The Manley–Rowe relations are:

$$\frac{d\Phi_1}{dz} = \frac{d\Phi_2}{dz} = -\frac{d\Phi_3}{dz}$$

For every created or annihilated photon at ω_3 there is one annihilated or created both at ω_1 and ω_2.

It is instructive to introduce modified field variables to describe the TWM interaction:

$$\tilde{A}_i = \sqrt{\frac{n_i}{\omega_i}}\tilde{E}_i \quad i = (1,2,3)$$

With these, the coupled wave equations become simpler:

$$\frac{\partial \tilde{A}_1}{\partial z} = -i\eta\tilde{A}_3\tilde{A}_2^* e^{-i\Delta kz}$$

$$\frac{\partial \tilde{A}_2}{\partial z} = -i\eta\tilde{A}_3\tilde{A}_1^* e^{-i\Delta kz}$$

$$\frac{\partial \tilde{A}_3}{\partial z} = -i\eta\tilde{A}_1\tilde{A}_2^* e^{-i\Delta kz}$$

The advantage is that the same coupling coefficient,

$$\eta = \sqrt{\frac{\omega_3\omega_2\omega_1}{n_3n_2n_1}\frac{\chi^{\mathrm{TWM}}}{2c}}$$

can be used in all three equations. $\left|\tilde{A}_i\right|^2$ is proportional to the photon current density: $\Phi_i = I_i / \hbar\omega_i = \left(c\varepsilon_0 / 2\hbar\right)\left|\tilde{A}_i\right|^2 \left(i = 1,2,3\right)$

The general term three-wave mixing includes a few specific processes: SFG, DFG, and optical parametric amplification (OPA). Essentially, these differ in the initial conditions.

Sum-frequency Generation (SFG)

Two intense input beams at frequencies ω_1 and ω_2 are used to generate a beam at the sum frequency $\omega_3 = \omega_2 + \omega_2$, where usually there is no input field. In the low conversion efficiency limit the amplitudes of the two input fields can be considered as constants, and a similar expression is obtained for the sum-frequency amplitude as in case of non-depleted-pump SHG:

$$\tilde{E}_3\left(z\right) = -i\frac{\omega_3}{2cn_3}\chi^{\mathrm{TWM}}\,\tilde{E}_1\,\tilde{E}_2\,z\sin c\left(\frac{\Delta kz}{2}\right)e^{i\Delta kz/2}$$

Difference-frequency Generation (DFG)

Two intense input beams at frequencies ω_3 and ω_2 are used to generate a beam at the difference frequency $\omega_1 = \omega_3 + \omega_2$, where usually there is no input field. In the low conversion efficiency limit the amplitudes of the two input fields can be considered as constants, and a similar expression is obtained for the sum-frequency amplitude:

$$\tilde{E}_1\left(z\right) = -i\frac{\omega_1}{2cn_1}\chi^{\mathrm{TWM}}\,\tilde{E}_3\,\tilde{E}_2^{*}\,z\sin c\left(\frac{\Delta kz}{2}\right)e^{-i\Delta kz/2}$$

We note that the case $\omega_1 \ll \omega_3, \omega_2$ and $\omega_3 \approx \omega_2$ is often referred to as optical rectification (OR) even though ω_1 is not strictly a DC field.

Optical Parametric Amplification (OPA)

TWM can also be used to amplify a weak input beam with the help of a strong pump beam. This situation is called optical parametric amplification (OPA). The pump beam has the highest frequency (ω_3). The beam to be amplified is called the signal; it can have any of the frequencies ω_1 or ω_2. The other of the two is called the idler.

In case of the non-depleted pump approximation $\tilde{A}_3$ is constant, while $\tilde{A}_1$ and $\tilde{A}_2$ are allowed to vary. The two relevant equations can be combined to give,

$$\frac{\partial^2 \tilde{A}_1}{\partial z^2} = g^2 \tilde{A}_{1-g\Delta k}\tilde{A}_2^{*} e^{-i\Delta kz}$$

$$\frac{\partial^2 \tilde{A}_2}{\partial z^2} = g^2 \tilde{A}_{2-g\Delta k}\tilde{A}_1^{*} e^{-i\Delta kz}$$

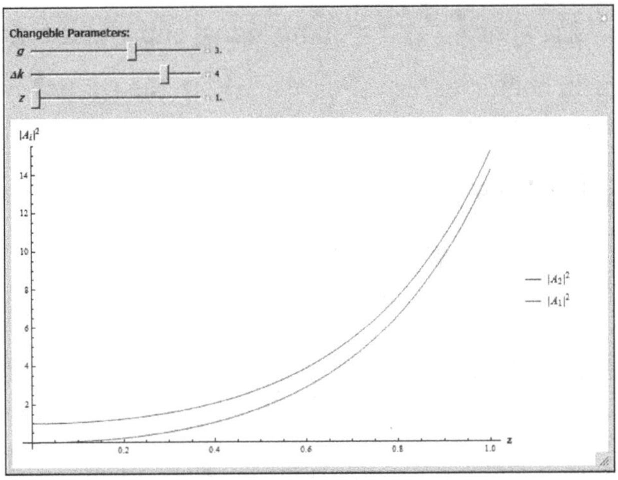

Evolution of signal and idler fields for $g = 3$, $\Delta k = 4$, and $\tilde{A}_2(0) = 1$. z is in units of $\pi / \Delta k$.

where g is the parametric gain coefficient given by,

$$g = \eta \tilde{A}_3 = \frac{\chi^{\mathrm{TWM}} \tilde{A}_3}{2\mathrm{c}} \sqrt{\frac{\omega_3 \omega_2 \omega_1}{n_3 n_2 n_1}}$$

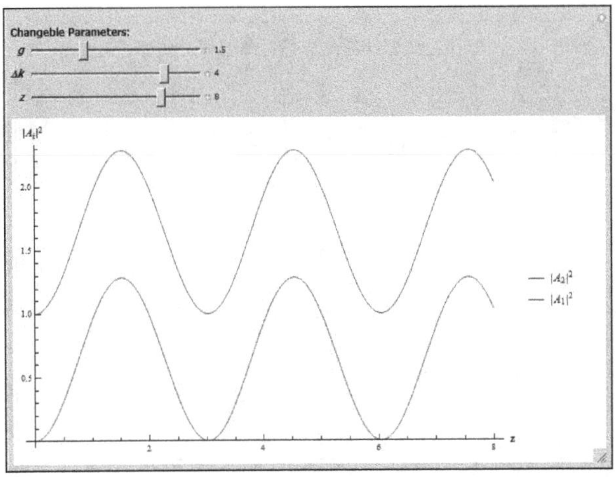

Evolution of signal and idler fields for $g = 1.5$, $\Delta k = 4$, and $\tilde{A}_2(0) = 1$. z is in units of $\pi / \Delta k$.

It can be assumed that g is real as long as the frequencies considered are far away from material resonances (the pump field can be considered real without any loss of generality). The above equatoins can be solved to give,

$$\tilde{A}_1(z) e^{i\Delta kz/2} = \tilde{A}_1(0) \cosh g'z + \mathrm{i}\left(\frac{1}{2}\Delta k \tilde{A}_1(0) - g\tilde{A}_2^*(0)\right)\left(\frac{\sinh g'z}{g'}\right)$$

$$\tilde{A}_2(z) e^{i\Delta kz/2} = \tilde{A}_2(0) \cosh g'z + \mathrm{i}\left(\frac{1}{2}\Delta k \tilde{A}_2(0) - g\tilde{A}_1^*(0)\right)\left(\frac{\sinh g'z}{g'}\right)$$

with $g' = \sqrt{g^2 - (\Delta k/2)^2}$. For $g > \Delta k/2$ both signal and idler fields grow exponentially with an effective gain coefficient g'. For $g < \Delta k/2$ the spatial evolution of the field amplitudes becomes oscillatory.

In case of zero initial idler field the solutions become:

$$\tilde{A}_1(z) = \tilde{A}_2^*(0) e^{-i\Delta kz/2} \left(-i \frac{g \sinh g'z}{g'} \right)$$

$$\tilde{A}_2(z) = \tilde{A}_2(0) e^{-i\Delta kz/2} \left(\cosh g'z + i \frac{\Delta k \sinh g'z}{2g'} \right)$$

In case of perfect phase matching the solutions have the form

$$\tilde{A}_1(z) = \tilde{A}_1(0) \text{coshgz} - i \tilde{A}_2(0) \text{sinhgz}$$
$$\tilde{A}_2(z) = \tilde{A}_2(0) \text{coshgz} - i \tilde{A}_1^*(0) \text{sinhgz}$$

which makes obvious that the initial phase relationship of the fields has a crucial effect on their evolution.

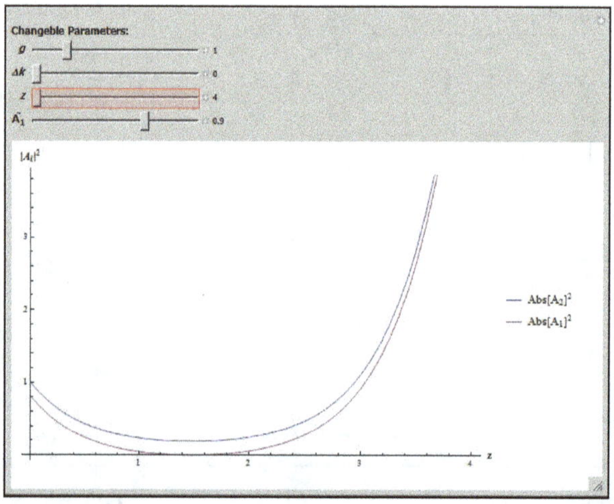

Evolution of signal and idler fields for $g = 1$, $\Delta k = 0$, $\tilde{A}_1(0) = 0.91$, and $\tilde{A}_2(0) = 1$. z is in units of $\pi/\Delta k$.

An interactive applet displaying the evolution of signal and idler fields is available here, where the parameters g, Δk, z, and $\tilde{A}_1(0)$ can be varied.

Optical Parametric Oscillators (OPOs)

The process of OPA can be used to build coherent light sources tunable over a wide range of frequencies. Such devices are ideal tools for many applications in spectroscopy.

They can be superior to tunable laser sources since they can have broader and smoother tuning properties, or they can access wavelength ranges where no broadband laser materials are available.

Tuning the wavelength of signal amplification in OPA can be accomplished by tuning phase matching to the required wavelength. This is possible in suitable birefringent nonlinear optical materials by using angle or temperature tuning. If phase matching is provided, a strong pump beam can generate radiation at signal and idler frequencies even in absence of external inputs at these frequencies. This originates from parametric fluorescence, a process having quantum mechanical origin. Building a resonator around such a pumped and phase-matched nonlinear crystal can enhance this low-level signal resulting in a coherent intense signal beam. The scheme of such an optical parametric oscillator (OPO) is shown in figure.

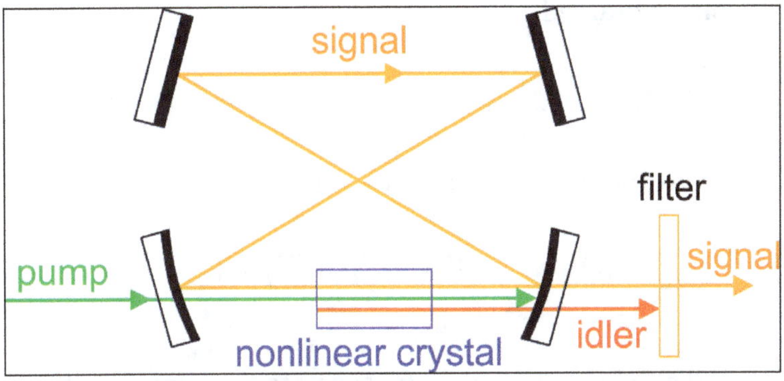

Scheme of an OPO singly resonant for the signal.

In the most general case third-order nonlinear processes can relate up to four different frequency components, $\omega_1, \omega_2, \omega_3, \omega_4$. The 3rd-order nonlinear susceptibility $\chi^{(3)}$ is a tensor of rank 4, described in the most general case by $3^4 = 81$ independent components. The 3rd-order nonlinear polarization can be given by following a similar procedure developed for the description of the 2nd-order effects. The polarization at frequency $\omega_4 = \omega_1 + \omega_2 + \omega_3$ is given by:

$$\hat{P}_i(\omega_4) = \frac{1}{4}\varepsilon_0 \sum_P \sum_{jkl} \chi^{(3)}_{ijkl}(\omega_1, \omega_2, \omega_3)\hat{E}_j(\omega_1)\hat{E}_k(\omega_2)\hat{E}_l(\omega_3)$$

The description of the general four-wave mixing (FWM) process involves the solution of four coupled wave equations.

These include third-harmonic generation (THG) and the Kerr effect. As a special case of the latter the intensity-dependent refractive index is considered.

Third-harmonic Generation

Third-harmonic generation (THG) starts with an input pump field at the fundamental frequency $\omega_1 = \omega$ and the 3rd-harmonic field is generated at the frequency $\omega_4 = 3\omega$.

The field inside the medium is composed of two monochromatic frequency components:

$$\mathrm{E}(z,t)=\frac{1}{2}\sum_{n=1,4}\hat{\mathrm{E}}_n(z)e^{i\omega_n t}+\text{c.c.}=\frac{1}{2}\sum_{n=1,4}\hat{\mathrm{E}}_n(z)e^{i(\omega_n t-k_n z)}+\text{c.c.}$$

The 3rd-order polarization can be given as follows:

$$\mathrm{P}^{(3)}(z,t)=\varepsilon_0\int_{-\infty}^{\infty}d\omega\int_{-\infty}^{\infty}d\omega'\int_{-\infty}^{\infty}d\omega''\chi^{(3)}(\omega,\omega',\omega'')\mathrm{E}(z,\omega)\mathrm{E}(z,\omega')\mathrm{E}(z,\omega'')e^{i(\omega+\omega'+\omega'')t}.$$

By substituting the Fourier amplitudes of the above two-component field into this expression and selecting the terms contributing to the polarization at the 3rd harmonic the following expression is obtained,

$$\hat{\mathrm{P}}_4^{(3)}(z)=\hat{\mathrm{P}}^{\mathrm{THG}}(z)=\frac{1}{4}\varepsilon_0\chi^{(3)}(\omega,\omega,\omega)\hat{\mathrm{E}}_1(z)\hat{\mathrm{E}}_1(z)\hat{\mathrm{E}}_1(z).$$

By writing out the coordinate indices explicitly this gives:

$$\hat{\mathrm{P}}_i(3\omega)\quad -\varepsilon\sum_{jkl}\chi_{ijkl}^{\mathrm{THG}}(\omega,\omega,\omega)\hat{\mathrm{E}}_j(\omega)\hat{\mathrm{E}}_k(\omega)\hat{\mathrm{E}}_l(\omega),$$

where i,j,k,l can take the values x,y,z. Please note the change in the meaning of the indices (denoting Cartesian coordinates rather than the frequency component) and the arguments of the field quantities (denoting the frequency rather than the propagation distance z). The z -dependence and the order of nonlinearity are not written out explicitly.

THG can be treated similarly to SHG, described by two coupled wave equations. In the limiting case of negligible pump depletion we get similar result for the 3rd harmonic field:

$$\tilde{\mathrm{E}}_4(z)=\mathrm{i}\frac{3\omega}{8cn_4}\chi^{\mathrm{THG}}\,\tilde{\mathrm{E}}_1\,\tilde{\mathrm{E}}_1\,\tilde{\mathrm{E}}_1\,z\,\mathrm{sinc}\left(\frac{\Delta kz}{2}\right)e^{i\Delta kz/2}$$

The phase-mismatch parameter is defined as $\Delta k=k_4-\ k_1$. The intensity of the third harmonic beam is given by:

$$I_4(z)=\frac{1}{2}c\varepsilon_0 n_4\left|\tilde{\mathrm{E}}_4(z)\right|^2=\frac{\left(3\omega\chi^{\mathrm{THG}}\right)^2}{4\varepsilon_0^2c^4n_4n_1^3}I_1^3z^2\,\mathrm{sinc}^2\left(\frac{\Delta kz}{2}\right)$$

Notice the cubic dependence of the THG intensity on that of the fundamental.

Kerr Effect

The Kerr effect describes the change of the refractive index of a medium induced by an external field via third-order nonlinear interaction. In the DC Kerr effect a strong DC field induces the refractive index change. In the optical or AC Kerr effect the refractive index change is induced by an optical field. A special case of this is when an intense optical beam modifies its own refractive index.

In this Section structurally isotropic media will be assumed for simplicity. Gases, liquids, and amorphous solids are examples of such media. Please note that cubic crystals are optically isotropic but not structurally. For structurally isotropic media only 21 out of 81 elements of $\chi^{(3)}$ are non-zero, which can be grouped in four types $\chi_1, \chi_2, \chi_3, \chi_4$. The following relations hold:

$$\chi_1 = \chi_2 + \chi_3 + \chi_4,$$

and for the individual tensor components of each type:

$$\text{type } 1: xxxx = yyyy = zzzz\left(= xxyy + xyxy + xyyx\right)$$
$$\text{type } 2: xxyy = yyzz = zzxx = yyxx = zzyy = xxzz$$
$$\text{type } 3: xyxy = yzyz = zxzx = yxyx = zyzy = xzxz$$
$$\text{type } 4: xyyx = yzzy = zxxz = yxxy = zyyz = xzzx$$

Hence, a structurally isotropic medium has 3 independent $\chi^{(3)}$ components.

DC Kerr Effect

A DC (zero-frequency) field $\hat{\mathrm{E}} = \mathrm{E}$ is applied across the medium through which propagates an optical field with frequency ω. The nonlinear polarization is:

$$\hat{P}_i(\omega) = 3\varepsilon_0 \sum_{jkl} \chi^K_{ijkl}(0,0,\omega) E_j(0) E_k(0) \hat{E}_l(\omega)$$

Let us assume that the DC field is in the y direction and the optical field propagates in the z direction. The nonlinear polarization at the optical frequency can be written as:

$$\hat{P}_x(\omega) = 3\varepsilon_0 \chi^K_{xyyx}(0,0,\omega) E_y^2(0) \hat{E}_x(\omega) = 3\varepsilon_0 \chi_4^K E_y^2(0) \hat{E}_x(\omega)$$
$$\hat{P}_y(\omega) = 3\varepsilon_0 \chi^K_{yyyy}(0,0,\omega) E_y^2(0) \hat{E}_x(\omega) = 3\varepsilon_0 \chi_1^K E_y^2(0) \hat{E}_y(\omega)$$

Thus, there is a nonlinear contribution to the polarization at the optical frequency which is proportional to the square of the amplitude of the applied DC field.

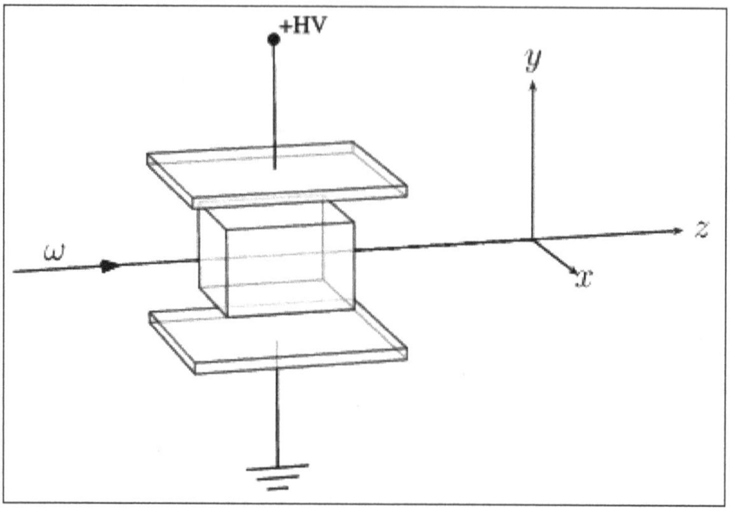

Arrangement for the DC Kerr effect.

The induced change of the refractive index is different for light polarized parallel and perpendicular to the applied DC field. This means that the birefringence of the medium is changed (or a birefringence is induced) as

$$\Delta n = n_{\parallel} - n\perp \simeq \frac{3\left(\chi_1^K - \chi_4^K\right)E_y^2(0)}{2n} = \frac{3\chi_2^K E_y^2(0)}{n}$$

A Kerr constant K is defined by the relation:

$$n_{\parallel} - n\perp = \lambda_0 K E^2(0)$$

with $\lambda_0 = 2\pi c/\omega$. The DC Kerr effect can be used for optical switching. This requires that the optical beam propagates a half-wave distance through the medium placed between crossed polarizers:

$$L\pi = \frac{\lambda_0}{2|\Delta n|} \simeq \frac{1}{2|K|E^2(0)}$$

Optical Kerr Effect

In the optical or AC Kerr effect an intense optical beam at frequency ω_2 modulates the refractive index for a co-propagating weak probe beam at frequency ω_1. This effect is also called as cross-phase modulation (XPM). Let us assume that both fields are polarized in the x direction. The nonlinear polarization is:

$$\hat{\mathrm{P}}_x(\omega_1) = \frac{3}{2}\varepsilon_0 \chi_{xxxx}^{\mathrm{OK}}\left(\omega_2, -\omega_2, \omega_1\right)\left|\hat{E}_x(\omega_2)\right|^2 \hat{E}_x(\omega_1)$$

The refractive index of the weak wave is changed by:

$$\Delta n_x \simeq \frac{3\chi_{xxxx}^{\mathrm{OK}} I(\omega_2)}{2n(\omega_1)n(\omega_2)c\varepsilon_0}$$

In case of x -polarized probe beam and y -polarized strong beam, the type 4 coefficient becomes effective rather than the type 1:

$$\hat{P}_x(\omega_1) = \frac{3}{2}\varepsilon_0\chi_{xyyx}^{\mathrm{OK}}(\omega_1;\omega_2,-\omega_2,\omega_1)\left|\hat{E}_y(\omega_2)\right|^2 \hat{E}_y(\omega_1)$$

Therefore, the index change is smaller.

Intensity-Dependent Refractive Index

Intensity-dependent refractive index (IDRI) is a special case of the optical Kerr effect. It occurs when an intense optical beam propagates in a medium thereby changing its refractive index. This self-induced refractive index change is influencing the propagation of the beam.

The field inside the medium has the form,

$$\mathrm{E}(z,t) = \frac{1}{2}\hat{\mathrm{E}}(z)e^{i\omega t} + \mathrm{c.c.}$$

By substituting the corresponding Fourier amplitude into the general expression of the 3rd-order polarization given above one obtains,

$$\hat{\mathrm{P}}_\omega^{(3)}(z) = \frac{3}{4}\varepsilon_0\chi^{(3)}(\omega,-\omega,\omega)\hat{\mathrm{E}}(z)\hat{\mathrm{E}}^*(z)\hat{\mathrm{E}}(z)$$

where intrinsic permutation symmetry (IPS) was taken into account. For a beam polarized in the x direction the nonlinear polarization is given by:

$$\hat{\mathrm{P}}_x(\omega) = \frac{3}{4}\varepsilon_0\chi_1^{\mathrm{IDRI}}\left|\hat{E}_x(\omega)\right|^2 \hat{E}_x(\omega)$$

The refractive index if the medium is:

$$n = n_0 + \left(\frac{3\chi_1^{\mathrm{IDRI}}}{4n_0^2 c\varepsilon_0}\right)I = n_0 + n_2 I$$

where n_0 is the low-intensity refractive index and n_2 is called the nonlinear refractive index.

In case of circularly and elliptically polarized light the effects of IDRI become more complex.

$$\hat{P}_x = \frac{1}{4}\varepsilon_0\left(A\left(\hat{E}\hat{E}^*\right)\hat{E}_x + \frac{1}{2}B\left(\hat{E}\hat{E}\right)\hat{E}_x^*\right)$$

$$\hat{P}_y = \frac{1}{4}\varepsilon_0\left(A\left(\hat{E}\hat{E}^*\right)\hat{E}_y + \frac{1}{2}B\left(\hat{E}\hat{E}\right)\hat{E}_y^*\right)$$

with $A = 6\chi_2^{\text{IDRI}}$ and $B = 6\chi_3^{\text{IDRI}}$. For a beam propagating in the z direction $\hat{E}\hat{E} = \hat{E}_x\hat{E}_x + \hat{E}_y\hat{E}_y$. Since the two terms containing the coefficient B in the above expressions contain $\hat{E}_x^*$ and $\hat{E}_y^*$. This has the consequence that when circularly (e.g. counter-clockwise) polarized light enters the medium this term will lead to the generation of nonlinear polarization with the opposite circular (e.g. clockwise) polarization. This is shown by the following expression explicitly:

$$\hat{P}_x - i\hat{P}_y = \frac{1}{4}\left(A\left(\hat{E}.\hat{E}\right)\left(\hat{E}_x - i\hat{E}_y\right) + \frac{1}{2}B\left(\hat{E}.\hat{E}\right)\left(\hat{E}_x - i\hat{E}_y\right)\right)$$

As a consequence of this feature, for elliptically polarized light, the axes of the ellipse rotate during propagation. For linearly or circularly polarized light the polarization state remains unchanged. The above expression giving the refractive index change illustrates this for linearly polarized light: the expression does not contain the coefficient B. For circularly polarized light the corresponding equation is:

$$n = n_0 + \left(\frac{A}{4n_0^2 c\varepsilon_0}\right)I$$

Self-focusing

IDRI has important practical relevance. One consequence is self-focusing. Self-focusing of light is the process in which an intense beam modifies the refractive index of the medium such that the beam is caused to come to a focus within the material. Here we have assumed that n_2 is positive. As a result, the laser beam induces a refractive index variation within the material with a larger refractive index at the center of the beam than at its periphery. Thus the material acts as if it were a positive lens, causing the beam to come to a focus within the material.

Another self-action effect is the self-trapping of light. In this process a beam propagates with a constant diameter as a consequence of an exact balance between self-focusing and diffraction. Self-trapping can occur only if the power carried by the beam is exactly equal to the so-called critical power for self-trapping.

$$P_{\text{cr}} = \alpha\frac{\lambda^2}{4\pi n_0 n_2}$$

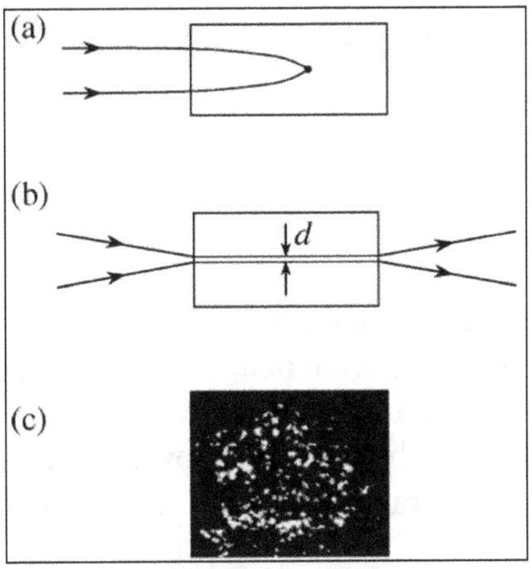

Schematic illustration of three self-action effects: (a) self-focusing of light, (b) self-trapping of light, and (c) laser beam breakup, showing the transverse distribution of intensity of a beam that has broken up into many filaments.

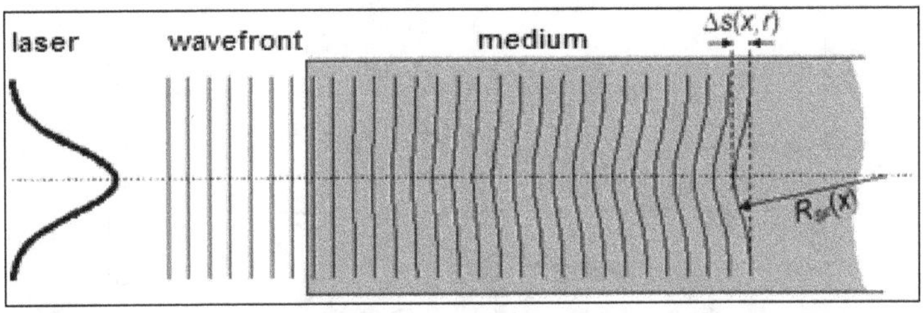

Distortion of the phase front of a collimated laser beam due to IDRI.

Self-focusing can occur only if the beam power P is greater than P_{cr}. The exact value of α depends on the precise beam profile, and is usually close to 2.

The final self-action effect shown in figure is laser beam breakup. This process occurs only for $P \gg P_{cr}$ and leads to the breakup of the beam into many components each carrying approximately power P_{cr}. This process occurs as a consequence of the growth of imperfections of the laser wavefront by means of the amplification associated with the forward four-wave mixing process.

A simple model of the self-focusing process can be given by ignoring the effects of diffraction. The neglect of diffraction is justified if the beam diameter or intensity (or both) is sufficiently large. Figure shows a collimated beam of characteristic radius ω_0 and an on-axis intensity I_0 falling onto a nonlinear optical material with positive n_2. Fermat's principle can be used to determine the distance z_{sf} from the input face to the self-focus. It states that the optical path length $\int n(r)\,dl$ of all rays traveling from a wavefront at the input face to the self-focus must be equal. As a first approximation, we take the refractive index along the marginal ray to be the linear refractive index n_0 and the refractive index along the central ray to be $n_0 + n_2 I_0$. According to Fermat's principle:

$$(n_0 + n_2 I_0) z_{sf} = \frac{n_0 z_{sf}}{\cos\theta_{sf}} \approx \frac{n_0 z_{sf}}{1 - \frac{1}{2}\theta_{sf}^2}$$

where the angle θ_{sf} is defined in the figure. Solving the approximate expression for θ_{sf}, we find that:

$$\theta_{sf} = \sqrt{\frac{2n_2 I_0}{n_0}}$$

This quantity is known as the self-focusing angle and in general can be interpreted as the characteristic angle through which a beam of light is deviated as a consequence of self-action effects. The ratio $n_2 I_0 / n_0$ of nonlinear to linear refractive index is invariably a small quantity, thus justifying the use of the paraxial approximation. In terms of the self-focusing angle, we can calculate the characteristic self-focusing distance as $z_{sf} = \omega_0 / \theta_{sf}$ or as:

$$z_{sf} = \omega\sqrt{\frac{n_0}{2n_2 I_0}} = \frac{2n_0\omega_0^2}{\lambda_0}\sqrt{\frac{P_{cr}}{P}} \left(\text{for } P \gg P_{cr}\right)$$

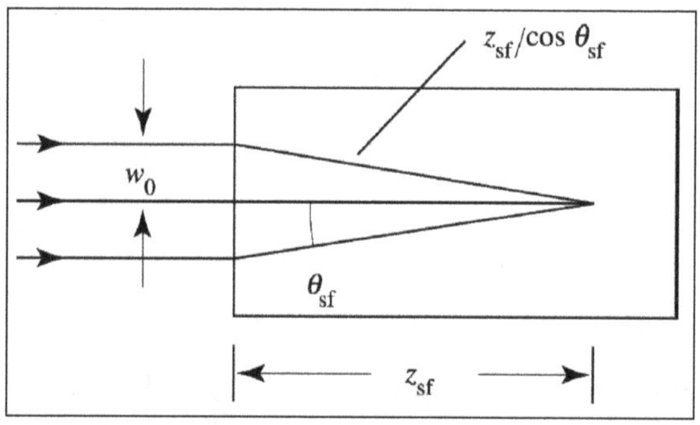

Prediction of the self-focusing distance z_{sf} by means of Fermat's principle. The curved ray trajectories within the nonlinear material are approximated as straight lines.

Anisotropic Media

In many cases, maybe in most cases, the nonlinear media used in practice are anisotropic. Anisotropy can play an essential role in achieving phase matching and in selecting large nonlinear susceptibilities for efficient frequency conversion processes. Therefore it is important to consider linear and nonlinear light propagation in anisotropic media. Linear and nonlinear optical properties are related to crystal structure and symmetries play an essential role.

Light Propagation in Linear Anisotropic Media

The most general form of the constitutive relation $D(E) = \varepsilon_0 E + P(E)$ for the case of linear response, which can be written as follows:

$$P(E) = \varepsilon_0 \chi^{(1)} E = P^{(1)}$$

$$D(E) = \varepsilon_0 \left(1 + \chi^{(1)}\right) E = \varepsilon_0 \varepsilon E$$

The linear susceptibility $\chi^{(1)}$ and the relative dielectric constant $\varepsilon = 1 + \chi^{(1)}$ are, in general, second-rank tensors, which can be represented by a 2-dimensional matrix having $3^2 = 9$ components. This is the case for anisotropic (non-isotropic) media. For linear, isotropic media, these quantities are scalars.

The second of the above equations can be written in matrix form as follows (the first relation is similar):

$$\begin{bmatrix} D_X \\ D_Y \\ D_Z \end{bmatrix} = \begin{bmatrix} \varepsilon_{XX} & \varepsilon_{XY} & \varepsilon_{XZ} \\ \varepsilon_{YX} & \varepsilon_{YY} & \varepsilon_{YZ} \\ \varepsilon_{ZX} & \varepsilon_{ZY} & \varepsilon_{ZZ} \end{bmatrix} \begin{bmatrix} E_X \\ E_Y \\ E_Z \end{bmatrix},$$

where the capital letters X, Y, Z denote an arbitrary orthogonal coordinate system. The key message of this equation is that D and E are not necessarily parallel in an anisotropic material. It is easy to show that the dielectric constant matrix is symmetric, and consequently it can be diagonalized by transforming it into the orthogonal coordinate system of the principal dielectric axes x, y, z:

$$\begin{bmatrix} D_x \\ D_y \\ D_z \end{bmatrix} = \begin{bmatrix} \varepsilon_{xx} & 0 & 0 \\ 0 & \varepsilon_{yy} & 0 \\ 0 & 0 & \varepsilon_{zz} \end{bmatrix} \begin{bmatrix} E_x \\ E_y \\ E_z \end{bmatrix},$$

The three diagonal elements are the three principal dielectric constants of the medium. Note that the principal axes are, in general, not the same as the crystallographic axes (a, b, c), which are not necessarily orthogonal.

It is obvious from the above Eq. that there are three different types of material. All three dielectric constants are different in biaxial media; two of the three are the same (by convention $\varepsilon_{xx} = \varepsilon_{yy}$) in uniaxial media; all three are the same in isotropic media.

In describing light propagation in a linear and anistropic medium one has to rely on Maxwell's equations and on the constitutive relation. Maxwell's equations impose constraints on the directions of the vectors E, D, B and H, which can easily be derived by considering plane waves of frequency ω propagating in the direction of the wave vector k.k is perpendicular to the wavefront. In substituting the space-time dependence $e^{i(\omega t - \mathrm{kr})}$ for all four field vectors into Maxwell's equations one obtains:

$$\mathrm{k\,D} = 0 \Rightarrow \mathrm{k} \perp \mathrm{D}$$

$$\mathrm{k\,B} = 0 \Rightarrow \mathrm{k} \perp \mathrm{B}$$

$$\mathrm{k} \times \mathrm{E} \propto \mathrm{B} \Rightarrow \mathrm{B} \perp \mathrm{k}, \mathrm{E}$$

$$\mathrm{k} \times \mathrm{B} \propto \mathrm{D} \Rightarrow \mathrm{D} \perp \mathrm{k}, \mathrm{B}$$

A geometric representation of these relations is given in figure. Since D and E are not necessarily parallel in an anisotropic material, so are k and the Poynting vector S. In such a case the direction of the energy flow differs from the wave propagation direction and an optical beam gets displaced along the wavefront as it propagates.

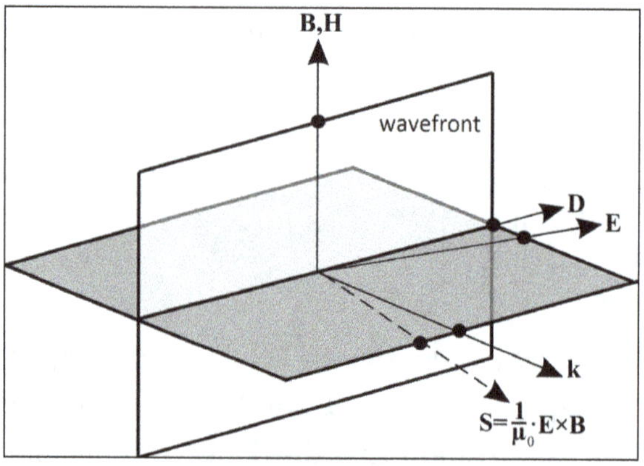

Directions of the wave vector, the field vectors and the Poynting vector for a plane wave in an anisotropic medium as they follow from Maxwell's equations.

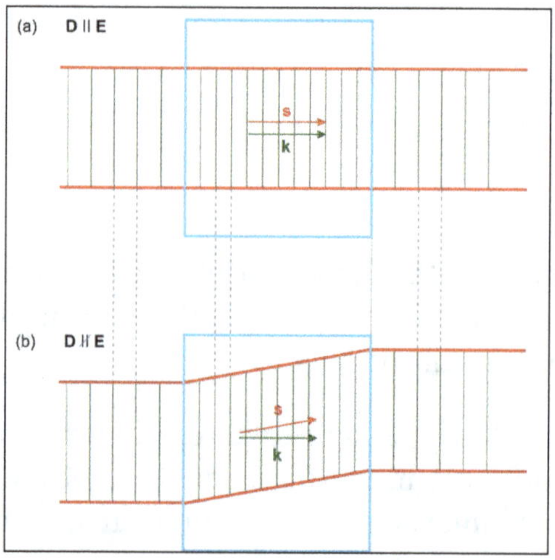

Plane-wave propagation in case of D and E being parallel (a) and non-parallel (b).

The symmetric 3×3 matrix ε determines a surface, called the index ellipsoid (or the wave normal ellipsoid), and is defined by the formula:

$$\frac{x^2}{\varepsilon_{xx}}+\frac{y^2}{\varepsilon_{yy}}+\frac{z^2}{\varepsilon_{zz}}=1,$$

or equivalently:

$$\frac{x^2}{n_x^2}+\frac{y^2}{n_y^2}+\frac{z^2}{n_z^2}=1,$$

where the corresponding refractive indices have been introduced. The index ellipsoid enables us to find the two phase velocities and the two directions of vibrations of D which belong to a given direction of k and which satisfy both the Maxwell's equations as well as the constitutive relation. We draw a plane through the origin at right angles to k. The curve of intersection of this plane with the ellipsoid is an ellipse. The semi-axes of this ellipse are proportional to the reciprocals of the phase velocities, and their directions coincide with the corresponding directions of vibrations of D.

In biaxial media, the three semi-axes of the ellipsoid n_x, n_y and n_z are all different. In uniaxial media $n_z = n_y \neq n_0$, where n_0 is the ordinary refractive index, and $n_z = n_e \neq n_0$, where n_e is the extraordinary refractive index. The x and y semi-axes are of equal length; the z -axis is known as the optic axis. If $n_e > n_0$, the medium is positive uniaxial; if $n_e < n_0$, the medium is negative uniaxial. The lengths of the semi-axes of the intersection ellipse (see the previous paragraph) are the refractive indices of the associated waves: n_0 for the ordinary wave, $n_e(\theta)$ for the extraordinary wave, where:

$$n_e(\theta) = \left(\frac{\cos^2\theta}{n_0^2} + \frac{\sin^2\theta}{n_e^2} \right)^{-\frac{1}{2}}$$

The consequence of D and E being non-parallel is that the direction of the energy flow differs from the wave propagation direction. The angle ρ between D and E, and equivalently between S and k, is the walk-off angle, given by:

$$\rho(\theta) = \pm \tan^{-1}\left[\left(\frac{n_0}{n_e} \right)^2 \tan \right] \mp \theta$$

where the upper and lower signs apply to negative and positive crystals, respectively. Positive ρ means that the Poynting vector points away from the optic axis. The walk-off angle is typically a few degrees. In frequency conversion processes, walk-off can limit the interaction length by limiting the region of spatial overlap between interacting beams.

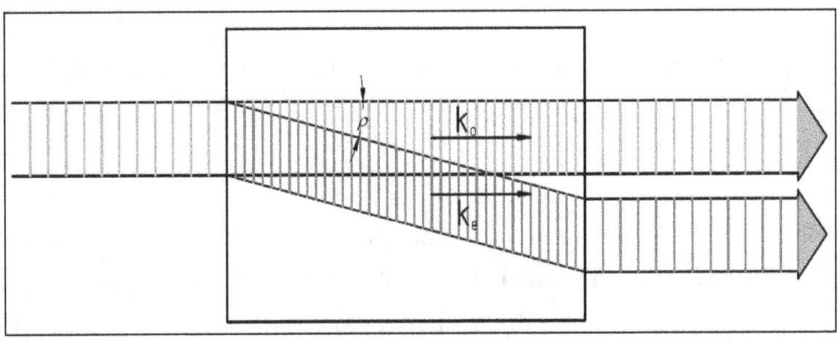

Walk-off between ordinary and extraordinary beams in a birefringent medium.

Nonlinear Optics in Anisotropic Media

The efficiency of nonlinear optical processes is determined by the nonlinear susceptibility of the medium and the phase mismatch parameter. The latter is the result of linear wave propagation. In order to choose the appropriate medium, and the propagation and polarization directions for the interacting fields, the detailed knowledge of the nonlinear susceptibility tensor is also needed. The second-order nonlinear susceptibility is a tensor of rank 3 with $3^3 = 27$ components depending on two independent field frequency components. The third-order nonlinear susceptibility is a tensor of rank 4 with $3^4 = 81$ components depending on three independent field frequency components, etc. However, various symmetry properties efficiently reduce the number of independent tensor components in practical situations. In the following a brief overview is given on the different symmetry properties.

Reality of the Fields

Owing to the reality of the electric fields and that of the nonlinear polarization the positive and negative frequency components of the susceptibility must be related according to:

$$\chi_{ijk}^{(2)}(-\omega_m, -\omega_n) = \chi_{ijk}^{(2)*}(\omega_m, \omega_n)$$

In words, changing the sign of both frequency arguments is equivalent to taking the complex conjugate.

Intrinsic Permutation Symmetry

For example, in case of SFG, the polarization amplitude at $\omega_3 = \omega_1 + \omega_2$ is given by:

$$\hat{P}_i(\omega_3) = \frac{1}{2}\varepsilon_0 \sum_{jk} \chi_{ijk}^{(2)}(\omega_1, \omega_2)\hat{E}_j(\omega_1)\hat{E}_k(\omega_2) + \frac{1}{2}\varepsilon_0 \sum_{jk} \chi_{ijk}^{(2)}(\omega_2, \omega_1)\hat{E}_j(\omega_2)\hat{E}_k(\omega_1)$$

Exchanging the order of fields, meaning the interchange of the terms $\chi_{ijk}^{(2)}(\omega_1, \omega_2)$ with $\chi_{ijk}^{(2)}(\omega_2, \omega_1)$, should not alter the expression. Therefore it is convenient to require that:

$$\chi_{ijk}^{(2)}(\omega_1, \omega_2) = \chi_{ijk}^{(2)}(\omega_2, \omega_1)$$

which is called intrinsic permutation symmetry. Note that it is the result of convenience.

Symmetries for Lossless Media

Two additional symmetries of the nonlinear susceptibility tensor occur for the case of a lossless nonlinear medium. The first of these conditions states that for a lossless medium all of the components of $\chi_{ijk}^{(2)}(\omega_1, \omega_2)$ are real. The general proof of this condition is obtained by verifying that the quantum-mechanical expression for $\chi^{(2)}$ is purely real in this limit.

The second of these symmetries is full permutation symmetry. This condition states that the frequency arguments of the nonlinear susceptibility can be freely interchanged, as long as the corresponding Cartesian indices are interchanged simultaneously. When applying this symmetry, it can be helpful to consider the susceptibility as being the function of all three frequencies involved in the interaction. For example one can write $\chi_{ijk}^{(2)}(\omega_3=\omega_1+\omega_2;\omega_1,\omega_2)=\chi_{ijk}^{(2)}(\omega_1,\omega_2)$. (Since the first argument is always the sum of the latter two, and therefore it is redundant, we avoid to use this notation elsewhere. In permuting the frequency arguments, the signs of the frequencies must be inverted when the first frequency is interchanged with either of the latter two. Full permutation symmetry implies, for instance, that:

$$\chi_{ijk}^{(2)}(\omega_1,\omega_2)=\chi_{ijk}^{(2)}(\omega_3,-\omega_2)=\chi_{ijk}^{(2)}(\omega_3,-\omega_1)$$

The importance of this symmetry condition is that it allows, when fulfilled, to use the same susceptibility tensor elements in the coupled wave equations describing three-wave interactions.

Kleinman's Symmetry

Additional symmetries apply in the case when all frequencies involved are much lower than the lowest resonance frequency of the medium. In this case the nonlinear susceptibility is essentially independent of the frequency. Full permutation symmetry also applies. Therefore, it is possible to permute the indices without permuting the frequencies:

$$\chi_{ijk}^{(2)}(\omega_1,\omega_2)=\chi_{jki}^{(2)}(\omega_1,\omega_2)=\chi_{kij}^{(2)}(\omega_1,\omega_2)=\chi_{jik}^{(2)}(\omega_1,\omega_2)=\chi_{ikj}^{(2)}(\omega_1,\omega_2)=\chi_{kji}^{(2)}(\omega_1,\omega_2)$$

This is the Kleinman symmetry condition.

Phase Matching

The efficiency of nonlinear optical processes is determined by the nonlinear susceptibility of the medium and the phase mismatch parameter. The latter is the result of linear wave propagation. To achieve optimal energy conversion efficiency, the phase-matching condition $\Delta k=0$ has to be fulfilled in many (but not all) nonlinear optical processes. This can be formulated as a linear relation between the wave vectors of the various frequency components involved in the process.

$$\Delta \mathrm{k}=\mathrm{k}_2-2\mathrm{k}_1=0 \text{ for SHG},$$
$$\Delta \mathrm{k}=\mathrm{k}_3-\mathrm{k}_1-\mathrm{k}_2=0 \text{ for SFG},$$
$$\Delta \mathrm{k}=\mathrm{k}_3-3\mathrm{k}_1=0 \text{ for THG}.$$

These relations can be translated into relations between the refractive indices at the respective frequencies. For example, in case of SHG, phase matching requires $n(\omega) = n(2\omega)$. Except for very specific cases, material dispersion usually does not allow to fulfil such kind of relations. The most commonly used technique of phase matching relies on utilizing birefringence.

Birefringent Phase Matching

In order to achieve phase matching through the use of birefringent crystals, the highest-frequency wave $\omega_3 = \omega_1 + \omega_2$ is polarized in the direction that gives it the lower of the two possible refractive indices. For the case of a negative uniaxial crystal this choice corresponds to the extraordinary polarization. There are two choices for the polarizations of the lower-frequency waves. Type I phase matching is defined to be the case in which the two lower-frequency waves have the same polarization, and type II to be the case where the polarizations are orthogonal. No assumptions regarding the relative sizes of ω_1 and ω_2 are implied by the classification scheme. However, for type II phase matching it is easier to achieve the phase-matching condition (i.e., less birefringence is required) if $\omega_2 > \omega_1$. Also, independent of the relative values of ω_1 and ω_2, type I phase matching is easier to achieve than type II.

Careful control of the refractive indices at each of the three optical frequencies is required in order to achieve the phase-matching condition. Typically phase matching is accomplished by one of two methods: angle tuning and temperature tuning.

Angle Tuning

This method involves precise angular orientation of the crystal with respect to the propagation direction of the incident light and properly selecting the polarizations of the interacting beams. Appropriate choice of polarizations selects between ordinary and extraordinary propagations.

Temperature Tuning

There is one serious drawback to the use of angle tuning. Whenever the angle θ between the propagation direction and the optic axis has a value other than 0 or 90 degrees, the Poynting vector S and the propagation vector k are not parallel for extraordinary rays. As a result, ordinary and extraordinary rays with parallel propagation vectors quickly diverge from one another as they propagate through the crystal. This walk-off effect limits the spatial overlap of the two waves and decreases the efficiency of any nonlinear mixing process involving such waves.

For some crystals, e.g. lithium niobate, the amount of birefringence is strongly temperature-dependent. As a result, it is possible to phase-match the mixing process by holding θ fixed at 90 degrees and varying the temperature of the crystal.

Quasi-phase-matching

In case $\Delta k = 0$ cannot be provided by other phase matching techniques, there is still a possibility to circumvent the limitation of the coherence length and to ensure the growth of the desired field amplitude beyond this length scale. Quasi-phase-matching provides a tool for this. It relies on allowing to grow the field amplitude over one coherence length, and then, before the field would start to decrease again, the relative phases of the interacting fields is manipulated such that the desired amplitude grows again when propagating over the next coherence length.

One possible realization of such a scheme is to invert the sign of the nonlinear susceptibility after each coherence length. Another possibility is to include, after each coherence length of nonlinear interaction, sections with only linear propagation which rearrange the relative phases.

Broadband Phase Matching

In designing frequency converters or parametric amplifiers for tunable or broadband radiation (e.g. ultrashort laser pulses) the main challenge is to provide sufficient bandwidth while simultaneously optimizing the conversion efficiency or the amplification gain. To reduce the length of the nonlinear medium is in many cases a possibility to increase the bandwidth. However, this is usually accompanied by the serious reduction of conversion efficiency or amplification gain.

Introducing additional degrees of freedom into the phase matching process can help. One possibility is to use non-collinear (vector) phase matching.

Measurement of Nonlinear Refraction and Absorption Photorefraction Optical Damage

Nonlinear Refraction and Absorption

Nonlinear refraction of materials can be explained by the third order term in the series of expansion of the polarization $P = \varepsilon_0 \left(\chi^{(1)} + \chi^{(2)} E + \chi^{(3)} E^2 + \cdots \right) E$, where E is the electric field strength. According to this the intensity dependent index of refraction can be given as:

$$n = n_0 + n_2 I,$$

where $I \sim E^2$ is the intensity, n_0 is the linear, n_2 is the nonlinear index of refraction. This type of nonlinearity is called the Kerr-effect.

When a light beam with intensity of $I(r,z)$ passes through a material having nonlinear index of refraction, due to the Kerr-effect the beam induces nonlinear phase shift as:

$$d\phi = n_2 I(r,z) k \, dz,$$

where k is the magnitude of the wavevector, $\mathrm{d}z$ is the propagation distance in the media. This effect is called self-phase modulation, that can be manifested is self focusing or defocusing. The intensity dependence of the index of refraction through the associated nonlinear phase shift has influence on the beam propagation in the media, hence giving the possibilities for n_2 measurement methods such as two beam time resolved interferometry, spectral analysis , degenerate four wave mixing measurements, beam distortion measurements , *P*-scan , and the *z*-scan. The *z*-scan method will be discussed in details in the following sections.

In nonlinear materials beside nonlinear refraction nonlinear absorption is also often presented described as:

$$\frac{dI}{dz} = -\beta I^2,$$

where β is the nonlinear absorption coefficient. By the *z*-scan method nonlinear absorption can be studied as well.

The Z-scan Technique

Because of its simplicity, accuracy and sensitivity *z*-scan is the most popular and powerful method for the characterization of the sign and magnitude of n_2, and the magnitude of β. Although several developments of the technique exist for different beam types and modes the method was originally elaborated for CW Gaussian single beam, present discussion also corresponds to this case. In the measurement the nonlinear sample is scanned along the propagation direction (*z*) of a focused Gaussian beam around its focal plane, and the transmittance is monitored in the far field versus the *z* position of the sample as shown in figure.

The nonlinear material behaves as a lens with variable focal length during the scanning. In case of positive n_2 the material behaves as a converging lens, so in prefocal positions decrease, in postfocal positions increase can be observed in the far field on axis intensity. The situation is opposite in case of samples having negative n_2. This is in agreement with curves shown in figure. In case of a usual measurement of n_2 (called "closed aperture" measurement in the following) the aperture size is significantly smaller than the far field beam radius, thus the transmitted power is proportional to the on-axis intensity. The transmittance curves are normalized with the transmittance value should be measured without nonlinear effects. Practically this value can be presented by placing the nonlinear sample far from the focus. The so-called *z*-scan curve is the normalized transmittance plotted versus the *z* position of the sample. In figure thin-sample *z*-scan curves with their typical valley-peak, and peak-valley structures can be seen for positive and negative nonlinearity, respectively.

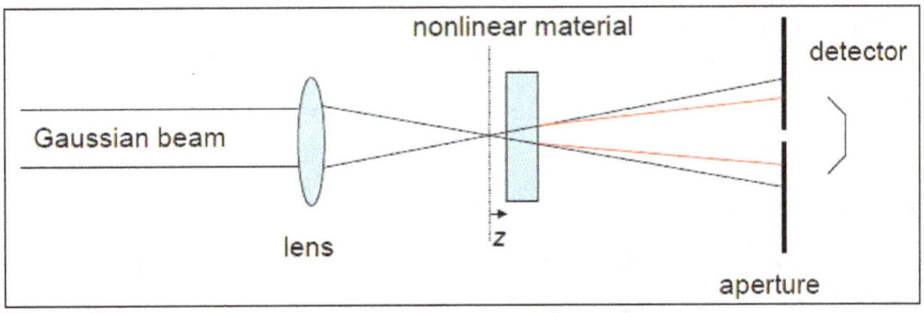

The z-scan measurement setup.

Depending on the thickness of the nonlinear media and on the magnitude of the nonlinearity different theoretical approaches exist for the evaluation of the measured curves. In the thin sample regime $L \ll z_0$ satisfies, where z_0 is the Rayleigh-range of the Gaussian beam, L is the sample length. The condition of weak nonlinearity satisfies, if $\Delta\phi_0 = n_2 I_0 kL \leq 0.2$, where I_0 is the on-axis intensity.

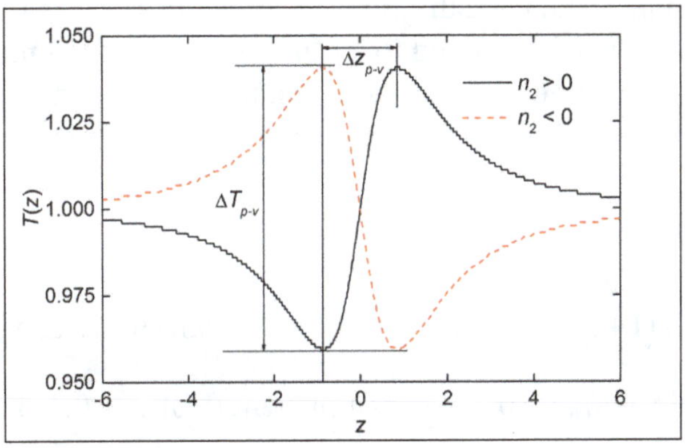

Z-scan curves of thin samples, having purely refractive nonlinearity.

Z-scan Study of Nonlinear Media

Thin Nonlinear Media

In the regime of weak optical nonlinearity by applying the so called Gaussian decomposition method in case of a purely nonlinear refractive sample one can deduce the following formula for the normalized transmittance:

$$T(z) = 1 - \frac{4\Delta\phi_0 \frac{z}{z_0}}{\left[\left(\frac{z}{z_0}\right)^2 + 9\right]\left[\left(\frac{z}{z_0}\right)^2 + 1\right]}$$

The illustrative curves in figure belong to laser parameters of $\lambda = 514\,\text{nm}, I_0 = 10^{10}\frac{\omega}{m^2}$ and $z_0 = 1\,\text{mm}$, and material parameters of $|n_2| = 1.6.10^{-14}\frac{\text{m}^2}{\text{W}}$, and $L = 0.1\,\text{mm}: |\Delta\phi_0| = 0.2$.

Please notice that with this very sensitive method refractive index changes with the order of $\Delta n = 10^{-5} - 10^{-4}$ can be examined. One can notice that the normalized transmittance value is $T = 1$ belonging to the $z = 0$ and $z = \pm\infty$, as it follows from the normalization condition in the latter case. Analyzing formula (8.4) it can be found that the peak-valley distance is $\Delta z_{p-v} \approx 1.7 z_0$, and the amplitude of the curve can be expressed as $\Delta T_{p-v} \approx 0.406|\Delta\phi_0|$. When measured curves are fitted with formula (8.4) if the Rayleigh-range is not known from other measurements, beside $n_2 z_0$ can be a fitting parameter too.

In case of thin nonlinear medium, with large $(\Delta\phi_0 < 0.2)$ nonlinearity there is no compact explicit formula for the theoretical *z*-scan curves. The algorithm has to be used is described in.

The most precise way in order to determine the nonlinear absorption is the "open aperture" *z*-scan technique. The measurement setup is the same as the "closed aperture" one, just the aperture has to be opened to let the whole beam to enter into the detector. In the regime of weak nonlinear absorption, when $\beta I_0 L \le 0.5$ the following analytical formula can be used for the open aperture normalized transmittance:

$$T(z) = \frac{1}{q_0(z)} \text{In}\left(1 + q_0(z)\right)$$

where $q_0(z) = \beta I_0 L / \left(1 + z^2 / z_0^2\right)$. Open aperture *z*-scan curves can be seen in figure beside constant intensity of $I_0 = 10^{10} \frac{\text{W}}{\text{m}^2}$, sample length of $L = 0.1\,\text{mm}$, and $z_0 = 0.1\,\text{mm}$. The varying β values are indicated in the figure. As a direct consequence of the nonlinear property of the absorption the transmission is minimal at $z = 0$.

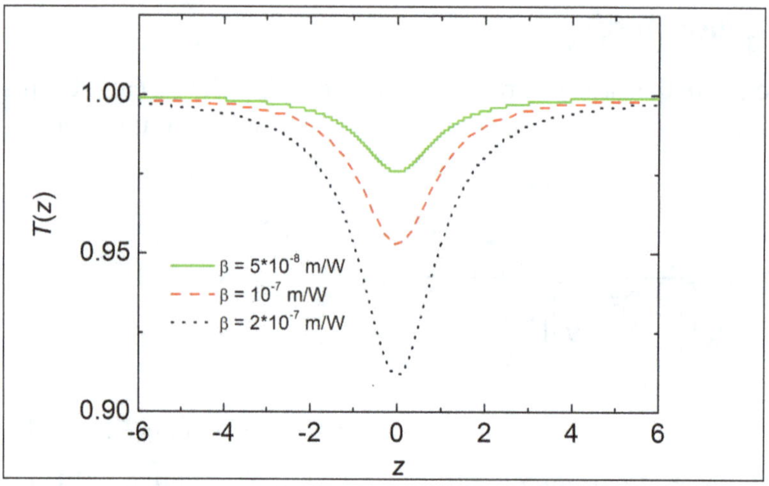

Open aperture *z*-scan curves of a thin sample, having weak nonlinear absorption.

Z-scan Study of Thick Nonlinear Media

If the $L \ll z_0$ condition does not satisfy. For thick media the closed aperture *z*-scan curves can be fitted by the following formula:

$$T(z)=1-\frac{1\Delta\phi_0}{4}z_0\,\mathrm{In}\left[\frac{9+\left(\frac{z-L}{z_0}\right)^2}{1+\left(\frac{z-L}{z_0}\right)^2}\cdot\frac{1+\left(\frac{z}{z_0}\right)^2}{9+\left(\frac{z}{z_0}\right)^2}\right]-\frac{1}{4}\beta I_0 z_0\left\{\tan^{-1}\left[\frac{4\frac{z}{z_0}}{3-\left(\frac{z}{z_0}\right)^2}\right]-\tan^{-1}\left[\frac{4\frac{z-L}{z_0}}{3-\left(\frac{z-L}{z_0}\right)^2}\right]\right\}.$$

This formula takes into account both nonlinear refraction and absorption, but it is valid only in the regime of weak nonlinearity. In figure theoretical *z*-scan curves belonging to parameters of $I_0=10^{10}\,\frac{\mathrm{W}}{\mathrm{m}^2}$, $L=2\,\mathrm{mm}$, $z_0=1\,\mathrm{mm}$ and $\Delta\phi_0=0.2$ can be seen. The absorptionless situation is plotted with solid black line. The effect of nonlinear absorption is also seen in the figure for various values of β. It is obvious that with increasing β absorption becomes the dominant effect, and the curve shape will be similar to the "open aperture" curves of figure.

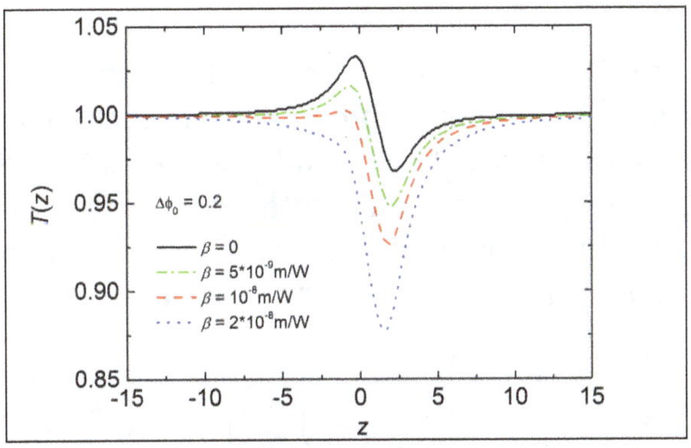

Z-scan curves of a thick sample with (green, red, blue lines) and without (black line) nonlinear absorption.

As an illustration for measurement evaluation figure shows the theoretical fitting with equation:

$$T(z)=1-\frac{1\Delta\phi_0}{4}z_0\,\mathrm{In}\left[\frac{9+\left(\frac{z-L}{z_0}\right)^2}{1+\left(\frac{z-L}{z_0}\right)^2}\cdot\frac{1+\left(\frac{z}{z_0}\right)^2}{9+\left(\frac{z}{z_0}\right)^2}\right]-\frac{1}{4}\beta I_0 z_0\left\{\tan^{-1}\left[\frac{4\frac{z}{z_0}}{3-\left(\frac{z}{z_0}\right)^2}\right]-\tan^{-1}\left[\frac{4\frac{z-L}{z_0}}{3-\left(\frac{z-L}{z_0}\right)^2}\right]\right\}.$$

The examined material is a well known nonlinear optical material: stoichiometric $\mathrm{LiNbO_3}$ with 5 mol% Mg content. As the configuration of the curve shows the material has positive nonlinear refractive index. The parameters of the laser are $I_0=2.10^9\,\frac{\mathrm{W}}{\mathrm{m}^2}$

, $z_0 = 1.7\,\text{mm}$, and the crystal thickness is $L = 2\,\text{mm}$. The obtained fitting values are

$n_2 = 6.6.10^{-15}\,\dfrac{\text{m}^2}{\text{W}}$, and $\beta = 1.7.10^{-8}\,\dfrac{\text{m}^2}{\text{W}}$.

For the case of arbitrarily large nonlinear refraction and absorption no compact fitting formula exists. In this case the normalized transmittance can be deduced by numerical algorithm.

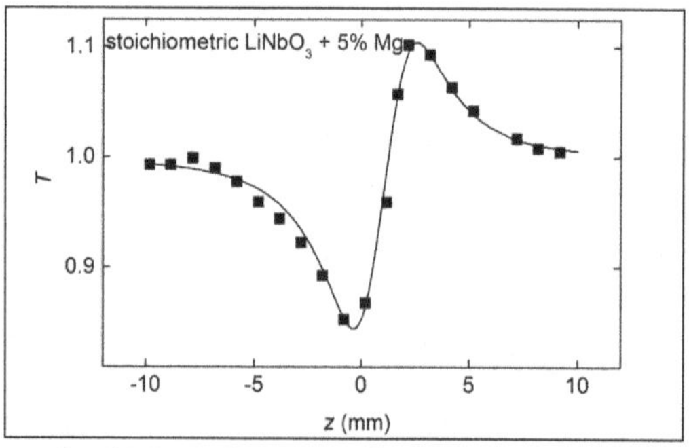

A measured z-scan curve of a thick $LiNbO_3$ crystal having nonlinear refraction and absorption.

General Theory of the Z-scan

The basis of the general theory of the *z*-scan is solving the wave equation by using paraxial approximation in a medium with nonlinear refraction and absorption. If the field amplitude is known at the exit of the sample the intensity in the far-field aperture zone is determined by the Huygens-Fresnel integral. By theoretical considerations of one can obtain for the so called nonlinear paraxial wave equation that describes the propagation in media having nonlinear refraction and absorption the following:

$$\frac{1}{\rho}\frac{\partial}{\partial\rho}\left(\rho\frac{\partial A}{\partial\rho}\right)-\left[\frac{2n}{n_0}|A|^2+\left(\frac{n_2^2}{n_0^2}-\frac{\beta^2}{4k^2n_0^2}\right)|A|^4\right]$$

$$k^2A-2ik\frac{\partial A}{\partial\xi}-ik\left(1+\frac{n_2}{n_0}|A|^2\right)\frac{\beta}{n_0}|A|^2A=0$$

where $\rho=\sqrt{x^2+y^2}$, ξ is the longitudinal coordinate, and *A* is the field amplitude ($I = A^2$). Please note, that for $n_2=0$ and $\beta=0$ the differential equation of the Gaussian beam is received back. The field amplitude A_{entrance} at the entrance of the sample positioned at $\xi = z$ can be determined by:

$$A_{\text{entrance}}(\rho,z)=\sqrt{I_0}\,\frac{\omega_0}{\omega(z)}\exp\left\{-\frac{\rho^2}{\omega(z)^2}+\mathrm{i}\left[\arctan\left(\frac{z}{z_0}\right)-\frac{k_0\rho^2}{2R(z)}\right]\right\}$$

as a Gaussian beam input. The intensity at the far-field aperture plane located at a distance $s \gg z_0$ from the focus can be determined by using the Huygens-Fresnel integral. According to this the field amplitude $A_{\text{aperture}}(r,z)$ in the aperture plane at a distance r from the optical axis can be determined as:

$$A_{\text{aperture}}(r,z) = \frac{\mathrm{i}}{\lambda} \int_0^{2\pi} \int_0^{\infty} \frac{A_{\text{exit}}(\rho, z+L) \exp\left(-\mathrm{i}k\sqrt{\rho^2 + r^2 + (s-z-L)^2 - 2\rho r \cos(\theta)}\right)}{\sqrt{\rho^2 + r^2 + (s-z-L)^2 - 2\rho r \cos(\theta)}} \rho d\rho d\theta.$$

The normalized transmittance through an aperture with a radius of r_a can be calculated as:

$$Tr_a(z) = \frac{\int_0^{r_a} |\text{Aaperture}(r,z)|^2 \, r dr}{\int_0^{r_a} |\text{Aaperture}(r, z=\infty)|^2 \, r dr}.$$

The often calculated normalized closed-aperture $(r_a \to 0)$ transmittance $T(z)$ is the following:

$$T(z) = \frac{|\text{Aaperture}(r=0,z)|^2}{|\text{Aaperture}(r=0,z=\infty)|^2}$$

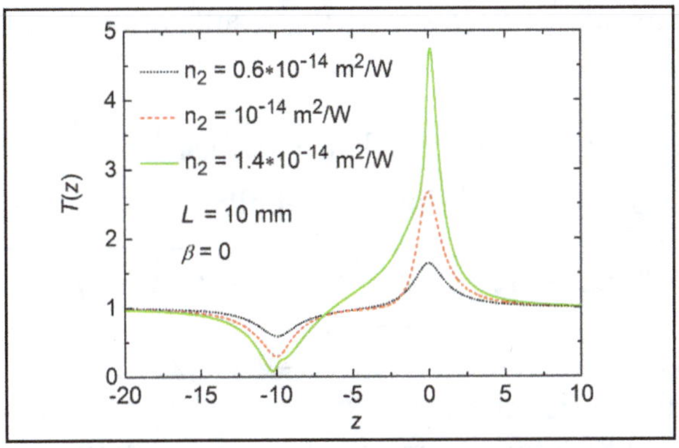

Z-scan results for thick samples $(L/z_0 = 10)$, and large nonlinearity.

As an illustration figure shows z-scan curves obtained by the algorithm described above for large sample thickness $L = 10\,\text{mm}$, strong nonlinearities without absorption. The laser parameters are $I_0 = 10^{10}\,\frac{\text{W}}{\text{m}^2}$, and $z_0 = 1\,\text{mm}$. The peak-valley distance for the curves in figure is approximately equal to the sample length. With increasing nonlinearity a fine structure appears around the minimum of the curve in figure.

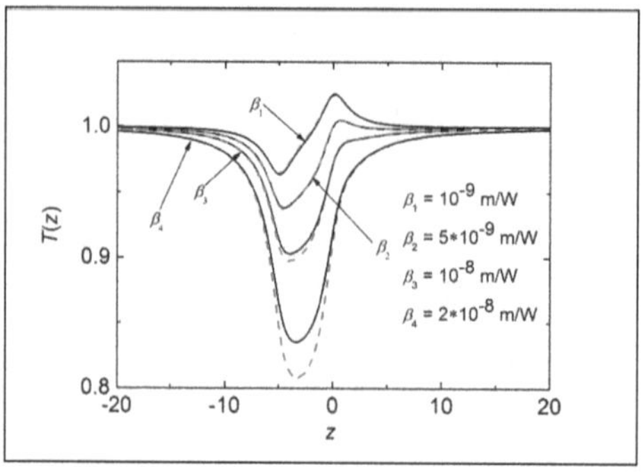

The effect of nonlinear absorption on the z-scan curves belonging to thick sample.

In figure the effect of nonlinear absorption on the transmittance can be seen obtained by the general theory of the z-scan. In the calculations $L = 5\,\text{mm}$ and a constant value of $n_2 = 5.10-16\frac{\text{m}^2}{\text{W}}$ was supposed and β was varied. The laser parameters were $I_0 = 10^{10}\frac{\text{W}}{\text{m}^2}$, and $z_0 = 1\,\text{mm}$. The results (black solid curves) are compared with the weak nonlinearity theory (red, dashed curves) given by

$$A_{\text{aperture}}(r,z) = \frac{\text{i}}{\lambda}\int_0^{2\pi}\int_0^{\infty}\frac{A_{\text{exit}}(\rho, z+L)\exp\left(-\text{i}k\sqrt{\rho^2+r^2+(s-z-L)^2-2\rho r\cos(\theta)}\right)}{\sqrt{\rho^2+r^2+(s-z-L)^2-2\rho r\cos(\theta)}}\rho d\rho d\theta.$$

For weak nonlinear absorption $(\beta I_0 L < 0.5)$ the results of the two theories are in very

good agreement. For $\beta I_0 L < 0.5$ no coincidence can be found between the two models, which means that $\beta I_0 L = 0.5$ is the limit of validity of equation.

$$A_{\text{aperture}}(r,z) = \frac{\text{i}}{\lambda}\int_0^{2\pi}\int_0^{\infty}\frac{A_{\text{exit}}(\rho, z+L)\exp\left(-\text{i}k\sqrt{\rho^2+r^2+(s-z-L)^2-2\rho r\cos(\theta)}\right)}{\sqrt{\rho^2+r^2+(s-z-L)^2-2\rho r\cos(\theta)}}\rho d\rho d\theta.$$

In the range of strong absorption the z-scan structure is dominated by the effect of nonlinear absorption, leading to suppression of the z-scan maximum.

The Photorefraction

Photorefraction means the following series of effects: when the material is exposed to light free charge carriers (electrons) are generated and they drifted from their original position leaving fixed charges (ionized donors) with opposite (positive) sign behind. The free carriers are trapped at other places. In case of nonuniform illumination the created space charge distribution is also nonuniform, and they create a nonuniform internal electric field. Due to the electro-optic effect this field induces a spatial dependent refractive index change that is proportional to the electric field. Switching off the illumination the "information" can be stored in the form of spatially alternating refractive index for a period time depending on the dark conductivity of the material. The information can be erased either by heating, or by illuminating with uniform light. In the followings interdependence will be deduced between the spatial dependent intensity of the illumination and the refractive index change using a simple one dimensional model applying several negotiations such as dark conductivity, volume photovoltaic effects, and the existence of holes.

- Generation of free charge holders (photoionization): The rate of generation of charge holders $G(x)$ is proportional to the $I(x)$ light intensity, and to the $N_D - N_D^+$ concentration difference between donors, and ionized donors as:

  $$G(x) = g\left(N_D - N_D^+\right)I(x).$$

  where g is the photoionization cross section.

- Recombination: The recombination rate $R(x)$ of the electrons is proportional to their concentration $n(x)$, and to the concentration of ionized donors (traps) as:

  $$R(x) = rn(x)N_D^+,$$

  where r is the recombination cross section. In equilibrium $R(x) = G(x)$ satisfies. Combining equation $G(x) = g\left(N_D - N_D^+\right)I(x)$ and $R(x) = rn(x)N_D^+$, one can obtain.

  $$n(x) = \frac{g}{r}\frac{N_D - N_D^+}{N_D^+}I(x),$$

  for the electron concentration.

- Current density and electric field strength: The resultant electric current density is the sum of the drift and diffusion current densities. In equilibrium it vanishes as:

  $$J(x) = e\mu n(x)E(x) - kT\mu\frac{dn}{dx} = 0$$

  where e is the elementary charge, μ is the electron mobility, k is the Boltzmann's constant, T is the temperature. Thus,

  $$E(x) = \frac{kT}{e}\frac{1}{n(x)}\frac{dn}{dx}.$$

- Refractive index change: According to the electro-optic effect:

  $$\Delta n(x) = -\frac{1}{2}n3r\,E(x),$$

where n and r are the appropriate value of the refractive index, and the electro-optic coefficient, respectively. Combining equations,

$J(x)=e\mu n(x)E(x)-kT\mu\frac{dn}{dx}=0$, $E(x)=\frac{kT}{e}\frac{1}{n(x)}\frac{dn}{dx}$. and

$\Delta n(x)=-\frac{1}{2}n3rE(x)$, one can obtain:

$$\Delta n(x)=-\frac{1}{2}n3r\frac{kT}{e}\frac{1}{I(x)}\frac{dI}{dx}$$

for the relation between the local refractive index change and the local intensity. This formalism can be generalized for two dimensions as well.

An important application of photorefraction is holographic storing. Frequently used photorefractive materials are $LiNbO_3$, $KNbO_3$, $Bi_{12}SiO_{20}$ and GaAs. Photorefractive sensitivity can be increased by adding different dopants to the material in during the crystal growth process. In case of $LiNbO_3$ a typical dopant is Fe. From the point of view of nonlinear optical applications the goal is to suppress photorefraction. For this typically Mg, Zn, In, Hf and Zr are used.

The Z-scan Study of Photorefraction

Since according to equation $\Delta n(x)=-\frac{1}{2}n3r\frac{kT}{e}\frac{1}{I(x)}\frac{dI}{dx}$ the light induced change of refractive index differs from the form of equation $n=n_0+n_2I$, the single beam z-scan technique can not be applied as quantitative method. Usually beam distortions during the scanning does not show circular symmetry as it is demonstrated with photorefractive congruent $LiNbO_3$ containing 5 mol % Mg shown in figure.

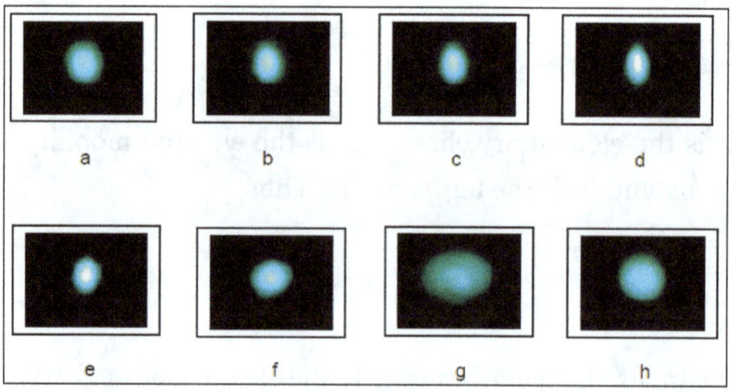

Beam distortions during the scanning of congruent $LiNbO_3$ containing 5 mol % Mg.

The corresponding z-scan curve is shown in figure.

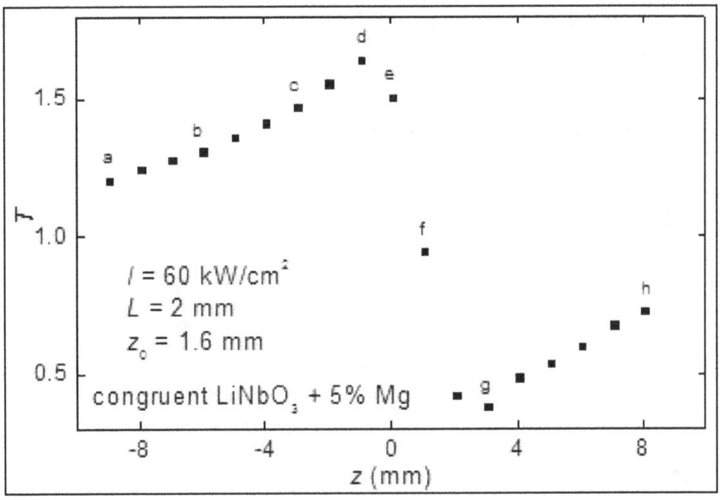

z-scan curve of congruent $LiNbO_3$ containing 5 mol % Mg.

As it is evident from figure and the asymmetry of the beam cross section is maximal in the *z*-scan maxima and minima. Furthermore it is seen, that the sign of the photorefractive nonlinearity is opposite as shown in figure for the non photorefractive nonlinear stoichiometric $LiNbO_3$ crystal.

Optical Damage

Optical damage of materials strongly limits applications. Two type of damages can be distinguished; if the damage occurs at the surface of the optical component we speak about surface damage, if the damage occurs internally in the bulk material we speak about inclusion damage. Generally the damage threshold is determined by the condition and quality of the surfaces rather than the bulk material. Intrinsic factors of the optical damage are absorption, color-center formation, nonlinear processes like self focusing, multiphoton absorption and electron avalanche breakdown. Extrinsic factors are impurities, contaminations, material defects, surface imperfections.

Surface Damage

One reason for the lower damage threshold of surfaces compared to the bulk is the residual defects of the surface even after an accurate polishing and cleaning process. Also minute inclusion of polishing material, or other impurities can be embed in the surface. Surface damage is caused either by absorption of submicrometer inclusions or by the formation of a plasma at the surface because of electron avalanche breakdown in the dielectric material.

In some cases application of dielectric coating of the surface increases the damage threshold by filling the scratches and micropits, making the surface smoother. Typical coating materials are MgF_2 on glass, ThF_4 on $LiNbO_3$, and sol-gel antireflection coating on fused silica. Instead of conventional polishing super-polishing techniques - such as ion polishing and chemical etching - lead to higher damage threshold.

Important to note that when a collimated beam is incident perpendicularly onto a plane-parallel dielectric material the damage occurs usually at the exit surface, not at the entrance. It can be explained by the behaviour of the Fresnel-reflection, since the intensity ratio at the exit to the entrance is:

$$\frac{I\,\text{ex}}{I\,\text{en}} = \left(\frac{2n}{n+1}\right)^2.$$

In order to prevent materials from the damage of the evanescent wave in case of internal reflective surfaces (Porro prisms, corner tubes) also surface coatings (typically SiO_2 with thickness of about 2 μm) are used.

Inclusion Damage

Minute inhomogeneities that are responsible for the inclusion damage generally come from the fabrication process of the material. These inhomogeneities can absorb light more efficiently giving the rise to local heating and melting of the material in that region. This in turn causes stress concentrations might be sufficient to rupture the material. Based on simple estimations it is found that the temperature of metal impurity particles with typical sizes can exceed 10000 K if the material is exposed to laser pulse with 20 J/cm^2 surface energy density and 30 ns pulse duration.

Pulse-width Dependence of the Damage Threshold

For pulses longer than about 100 ps damage is caused by conventional heat deposition mechanism resulting in melting and boiling. Series of experiments showed that the damage threshold scales with τ^{α}, where the typical value of α is between 0.3 and 0.6 in case of a variety of dielectric materials in the 20 ps to 100 ns range. figure illustrates the pulse-width dependence of the damage threshold for a number of materials.

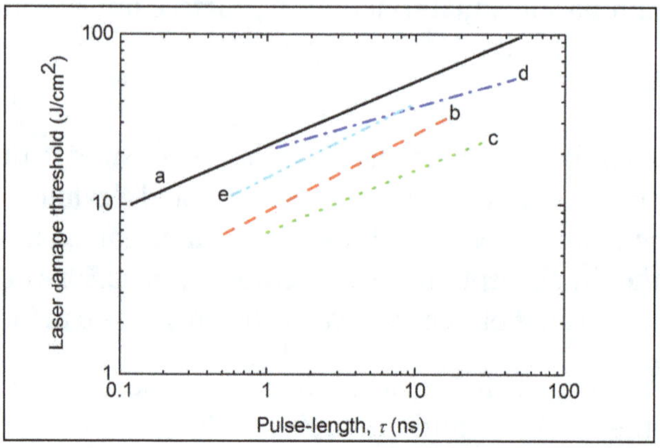

Damage threshold versus pulse-length for fused silica, BK-7, ULE, LG 750, and fluorophosphates glass at 1064 nm (a); fused silica at 355 nm (b); HfO/SiO_2 HR coating at 1064 nm, unconditioned (c) and laser conditioned (d); sol-gel AR coating on fused silica at 1064 nm and 355 nm (e).

The pulse-width dependence ranges from $\tau^{0.5}$ for fused silica at 355 nm and the sol-gel antireflection coating to $\tau^{0.4}$ for fused-silica, BK-7, ULE,LG 750, and fluorophosphates at 1064 nm, and to $\tau^{0.3}$ for the HfO_2 coating.

Decreasing the pulse-width below 100 ps, a gradual transition takes place from the long pulse, thermally dominated regime to an ablative regime dominated by collisional and multiphoton ionization and plasma formation. The damage threshold continues to decrease with decreasing pulse-width, but at a rate shorter than $\tau^{0.5}$.

Wavelength Dependence of the Damage Threshold

Typically, the damage threshold decreases for shorter wavelengths and drops off rather sharply at the shortest wavelengths as illustrated in figure for KDP.

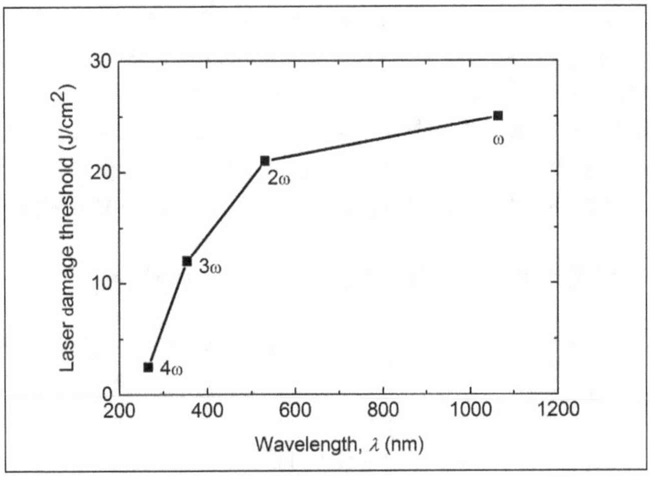

Bulk laser damage threshold of KDP crystal measured with 3 ns pulses at the fundamental mode and harmonics of a Nd:YAG laser.

For example for silica glass the laser-induced inclusion damage threshold can be given by the empirical formula: $J_{th} = 1.45\lambda^{0.43}$. J_{th} is the damage threshold in J/cm² and λ is the wavelength in nm.

Laser Host Materials and Nonlinear Crystals

The optical damage threshold of the laser material is rather determined by the coatings than the bulk material. Bulk damage threshold measurements with 10 ps Nd:YAG laser showed, that the lowest power density for which damage was observed was 3.6 GW/cm² for Nd:YAG sample, and 2.6 GW/cm² for Nd:YALO.

Among all optical materials the largest spread in damage thresholds occurs in the case of nonlinear materials. In some frequently used nonlinear materials are collected with their nominal damage threshold value. Please note the two orders of magnitude wideness of the range.

Table: Typical damage thresholds for nonlinear crystals. This table occurs in similar form.

Nonlinear crystal	KDP	KD*P	LBO	BBO	KTP KTA	$LiNbO_3$ PPLN	$AsGaS_2$ AsGaSe
Damage threshold (GW/cm^2)	8	4	2.5	1.5	0.5	0.2	0.03

Important to note that beside conventional effects responsible for optical damage in nonlinear crystals multiphoton absorption has important role.

Ultrashort light pulses play an important role in nonlinear optics. Since they provide a temporal concentration of light energy, the peak intensity can be many orders of magnitude higher than the intensity in a continuous-wave laser beam. This can lead to very pronounced nonlinear phenomena even with a moderate laser power. On the other hand, pulse propagation and interaction of pulses is influenced by dispersion, i.e. the frequency-dependence of the refractive index (or that of propagation direction in case of angular dispersion, or that of mode index in case of waveguides). In many cases this leads to additional complexity of nonlinear interactions and the wave equations.

Parametric Amplification with Ultrashort Pulses

In case of continuous laser beams the coupled wave equations governing OPA are in the following form:

$$\frac{\partial}{\partial z}\tilde{E}_1(z) = -i\frac{\omega_1}{2cn_1}\chi^{(2)}(\omega_3, -\omega_2)\tilde{E}_3(z)\tilde{E}_2^*(z)e^{-i\Delta kz}$$

$$\frac{\partial}{\partial z}\tilde{E}_2(z) = -i\frac{\omega_2}{2cn_2}\chi^{(2)}(\omega_3, -\omega_1)\tilde{E}_3(z)\tilde{E}_1^*(z)e^{-i\Delta kz}$$

$$\frac{\partial}{\partial z}\tilde{E}_3(z) = -i\frac{\omega_3}{2cn_3}\chi^{(2)}(\omega_1, \omega_2)\tilde{E}_1(z)\tilde{E}_2(z)e^{-i\Delta kz}$$

These equations need to be modified to account for propagation effects specific for short pulses. For simplicity, we assume that the idler (i), signal (s), and pump (p) fields all propagate in the z-direction, are linearly polarized and only one of their field components is non-zero. In this case we can drop the vector notation and we use i, s and p indices. However, one has to take into account that the fields are not monochromatic any more, which means that the slowly-varying field amplitudes become time-dependent:

$$E(z,t) = \frac{1}{2}\tilde{E}(z,t)e^{i(\omega t - kz)} + \text{c.c.}$$

propagating in the nonlinear crystal with different group velocities $v_g = \frac{d\omega}{dk}$.

By using slowly varying amplitudes, neglecting pulse lengthening and third-order non-linear effects (self- and cross-phase modulation), the coupled wave equations governing OPA of ultrashort pulses can be written as follows:

$$\frac{\partial A_s}{\partial z} + \frac{1}{v_{gs}} \frac{\partial A_s}{\partial t} = -\frac{i(\omega_s d_e \text{ff})}{n_s c_0} A_i^* A_p e^{-i\Delta kz}$$

$$\frac{\partial A_i}{\partial z} + \frac{1}{v_{gi}} \frac{\partial A_i}{\partial t} = -\frac{i(\omega_i d_e \text{ff})}{n_i c_0} A_s^* A_p e^{-i\Delta kz}$$

$$\frac{\partial A_p}{\partial z} + \frac{1}{v_{gp}} \frac{\partial A_p}{\partial t} = -\frac{i(\omega_p d_e \text{ff})}{n_p c_0} A_s A_i e^{-i\Delta kz}$$

By transforming to a frame of reference that is moving with the group velocity of the pump pulse $\tau = t - \frac{z}{v_{gp}}$ we obtain the equations:

$$\frac{\partial A_s}{\partial z} + \frac{1}{v_{gs}} - \frac{1}{v_{gp}} \frac{\partial A_s}{\partial \tau} = -\frac{i(\omega_s d_e \text{ff})}{n_s c_0} A_i^* A_p e^{-i\Delta kz}$$

$$\frac{\partial A_i}{\partial z} + \frac{1}{v_{gi}} - \frac{1}{v_{gp}} \frac{\partial A_i}{\partial \tau} = -\frac{i(\omega_i d_e \text{ff})}{n_i c_0} A_s^* A_p e^{-i\Delta kz}$$

$$\frac{\partial A_p}{\partial z} = -\frac{i(\omega_p d_e \text{ff})}{n_p c_0} A_s A_i e^{-i\Delta kz}$$

Group-velocity mismatch (GVM) between the pump and the amplified pulses limits the interaction length over which parametric amplification takes place, while GVM between the signal and the idler beams limits the phase matching bandwidth.

The useful interaction length for parametric interaction is quantified by the pulse splitting length, which is defined as the propagation length after which the signal (or idler) pulse separates from the pump pulse in the absence of gain:

$$l_{jp} = \frac{\tau}{\delta_{jp}}, \quad j = s, i$$

where τ is the pump pulse duration and $\delta_{jp} = \frac{1}{v_{gj}} - \frac{1}{v_{gp}}$ is the GVM. GVM depends on the crystal type, pump wavelength, and type of phase matching.

When $\delta_{sp}\delta_{ip} > 0$, both the signal and the idler pulses walk away from the pump in the same direction so that the gain rapidly decreases for propagation distances longer than the pulse splitting length and eventually saturates. On the other hand, when $\delta_{sp}\delta_{ip} < 0$ signal and idler pulses move in opposite direction with respect to the pump; in this way the signal and idler pulses tend to stay localized under the pump pulse and the gain grows exponentially even for crystal lengths well in excess of the pulse splitting length.

GVM between signal and idler pulses determines the phase matching bandwidth for the parametric amplification process. Assume perfect phase matching for a given signal frequency ω_s. If the signal frequency increases to $\omega_s + \Delta\omega$, by energy conservation the idler frequency decreases to $\omega_i - \Delta\omega$. The wave vector mismatch can then be approximated to the first order as:

$$\Delta k \simeq \frac{\partial \sim k_s}{\partial \omega s}\Delta\omega + \frac{\partial k_i}{\partial \omega_i}\Delta\omega = \left(\frac{1}{v_{gi}} - \frac{1}{v_{gs}}\right)\Delta\omega.$$

The FWHM phase matching bandwidth can then, within the large-gain approximation, be calculated as:

$$\Delta v \simeq \frac{\frac{2\sqrt{\ln 2}}{\pi}\sqrt{\frac{\Gamma}{L}}^{1}}{\left|\frac{1}{v_{gs}} - \frac{1}{v_{gi}}\right|}$$

Large GVM between signal and idler waves dramatically decreases the phase matching bandwidth; large gain bandwidth can be expected when the OPA approaches degeneracy $\omega_s \to \omega_i$ in type I phase matching or in the case of group velocity matching between signal and idler ($v_{gs} = v_{gi}$). In this case Δk must be expanded to the second order, giving:

$$\Delta v = \frac{2\sqrt[4]{in2}}{\pi}\sqrt[4]{\frac{\Gamma}{L}}\frac{1}{\left|\frac{\partial^2 k_s}{\partial \omega_s^2} + \frac{\partial^2 k_i}{\partial \omega_i^2}\right|}$$

Additional degree of freedom can be introduced using a non-collinear geometry, pump and signal wave vectors form an angle α and the idler is emitted at an angle Ω with respect to the signal. In this case the phase matching condition becomes a vector equation, which, projected on directions parallel and perpendicular to the signal wave vector, becomes:

$$\Delta k_{\|} = k_p \cos(\alpha) - k_s - k_i \cos(\Omega) = 0$$

$$\Delta k_{\perp} = k_p \sin(\alpha) - k_i \sin(\Omega) = 0$$

Note that the angle Ω is not fixed, but depends on the signal wavelength. If the signal frequency increases by $\Delta\omega$, the idler frequency decreases by $\Delta\omega$ and the wave vector mismatches along the two directions can be approximated, to the first order, as:

$$\Delta k_{\parallel} \simeq -\frac{\partial k_s}{\partial \omega_s}\Delta\omega + \frac{\partial k_i}{\partial \omega_i}\cos(\Omega)\Delta\omega - k_i \sin(\Omega)\frac{\partial \Omega}{\partial \omega_i}\Delta\omega$$

$$\Delta k_{\perp} \simeq \sim -\frac{\partial k_i}{\partial \omega_i}\sin(\Omega)\Delta\omega + k_i \cos(\Omega)\frac{\partial \Omega}{\partial \omega_i}\Delta\omega$$

To achieve broadband phase matching, both $\Delta k_{\parallel}$ and $\Delta k_{\perp}$ must vanish. Upon multiplying equations, upper by $\cos(\Omega)$ and downer by $\sin(\Omega)$ and adding the results, we get:

$$\frac{\partial k_i}{\partial \omega_i} - \cos(\Omega)\frac{\partial k_s}{\partial \omega_s} = 0$$

which is equivalent to:

$$v_{gs} = v_{gi} \cdot \cos(\Omega)$$

This equation shows that broadband phase matching can be achieved for a signal-idler angle Ω such that the signal group velocity along the signal direction.

For collinear geometry, signal and idler moving with different group velocities get quickly separated giving rise to pulse lengthening and bandwidth reduction, while in the non-collinear case the two pulses manage to stay effectively overlapped. This equation can be satisfied only if $v_{gi} > v_{gs}$; this is always the case in the commonly used type I phase matching in negative uniaxial crystals, where both signal and idler see the ordinary refractive index. This equation allows to determine the signal-idler angle Ω required for broadband phase matching; from a practical point of view, it is more useful to know the pump-signal angle α, which is given by:

$$\alpha = \arcsin\left(\frac{1 - \frac{v_{gs}^2}{v_{gi}^2}}{1 + \frac{2 v_{gs} n_s \lambda_i}{v_{gi} n_i \lambda_s} + \frac{n_s^2 \lambda_i^2}{n_i^2 \lambda_s^2}}\right)^{\frac{1}{2}}$$

Nonlinear Fiber Optics

Fiber optics as a research area exists since the 1950s. Although the main principle of fiber optics, namely total internal reflection (TIR), was discovered (but not explained) by Kepler in the 17th century, and uncladded ("bare") glass fibers based on this phenomenon were produced in the 1920s (J.L. Baird, C.W. Hansell), the core-and-cladding design that showed completely new characteristics was first used only in the 1950s by van Heel.

Since then many aspects of light propagation in fibers were discovered and used for various devices. The aim in the 1960s was to transmit images through a bunch of fibers, in which case the main limiting factor was the loss in the individual fibers. This problem was solved in the 1970s when loss was reduced below 0.2 dB/km using a wavelength of 1550 nm; in that case Rayleigh-scattering was the limitation factor. These fibers could already be used for optical fiber communication.

The decade of 1970 was also the decade of recognizing the importance of nonlinear effects -like stimulated Raman- and Brillouin scattering (SRS, SBS), possibility of soliton-like pulses etc.- in fibers. Later, in the 1980s, nonlinear effects in fibers were used for pulse compression and optical switching. A change in the doping material of the fibers (using rare-earth elements) made it possible to create all-optical amplifiers, leading to a revolution in fiber-optic communication.

The new millennium saw the discovery of new types of fiber-optic amplifiers based on stimulated Raman scattering and four-wave mixing (FWM). The novelty of these amplifiers was that doping was unnecessary and these fibers could be used at any spectral region. With these new developments ultrafast signal processing became possible.

In the last decade new types of solitons such as dispersion-managed and dissipative solitons as well as photonic crystal fibers (PCF; also called holey fiber, hole-assisted fiber, microstructure fiber, or microstructured fiber) were developed. These fibers are characterized with a relatively narrow core and a cladding with air holes. This structure shows much larger nonlinear effects that can be observed in fibers of length in the cm range.

The above historical summary shows that the field of nonlinear fiber optics is a field of such diversity that a chapter can hardly cover this topic. Therefore in this book we only present the basic principles of the field. For further information the interested reader is referred to the books specialized on the topic.

Characterizing the Fibers

The first fibers created in the 1920s were uncladded fibers. The core-and-cladding design appeared only in the 1950s. The simplest form of this kind of fibers is called step-index fiber (upper part of figure) which consists of a central glass core of refractive index n_{in} greater than that of the surrounding material called cladding $n_{out} < n_{in}$. In a more complex design the refractive index increases gradually from the core and cladding boundary towards the center of the core; these fibers are called graded-index fibers (lower part of figure).

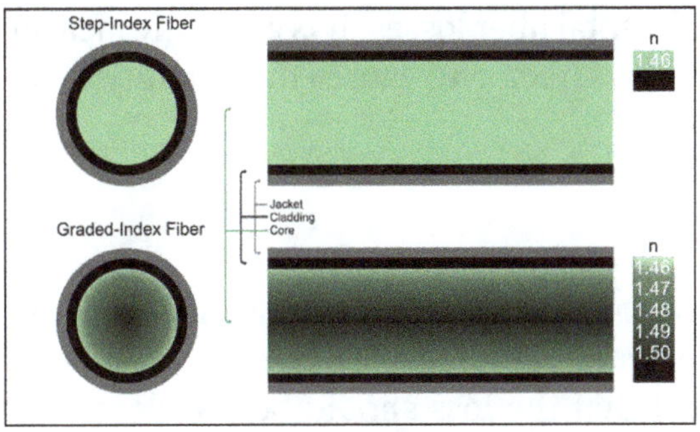

Step index and graded index fibers.

There are two parameters that are used to characterize the guiding properties of a fiber:

- V-parameter: This dimensionless parameter determines the number of modes supported by the fiber. For core radius α and wavelength λ this parameter can be expressed as:

  $$V = k_0 a\sqrt{n_{\text{in}}^2 - n_{\text{out}}^2}$$

  where $k_0 = 2\pi / \lambda$ is the wave number.

- Relative index difference: This quantity is expressed as:

  $$\Delta = \frac{n_{\text{in}} - n_{\text{out}}}{n_{\text{in}}}$$

If a fiber supports only one mode, it is called *single-mode fiber*. In the case of step-index fibers the single mode criterion is $V < 2.405$; for these fibers the core radius is typically $\alpha < 5\mu$m. Such fibers are mainly used in applications for optical communication. The V-parameter also shows that the difference between single-mode and multimode fibers is mainly in the core size a. One can also conclude that as far as the cladding is thick enough to support confinement, the radius of the cladding is not critical.

Fiber Losses

Attenuation, that is to say, intensity reduction with respect to distance traveled through a transmission medium, occurs in every fiber. To characterize it, first express the intensity ratio P_i/P_t as:

$$\frac{P_i}{P_t} = 10\alpha L$$

where L is the fiber length, P_t is the transmitted intensity and P_i is the initial (input) intensity. The proportionality factor α is called attenuation coefficient; this is the quantity that describes the total fiber losses. It is generally measured in dB/km, and from it can be expressed as follows (factor 10 causes the quantity to be measured in decibel):

$$\alpha = \frac{10}{L}\lg\left(\frac{P_i}{P_t}\right)$$

Losses originate from various phenomena, such as material absorption, Rayleigh scattering, scattering on the core-cladding interface, or even bending of the fiber. From this list one may suspect that the loss should be wavelength dependent. Indeed, loss spectrum can be assigned to the individual fibers. As an example, the Rayleigh scattering loss, which is a fundamental loss mechanism, varies with the wavelength as λ^{-4}, therefore it is dominant at lower wavelengths.

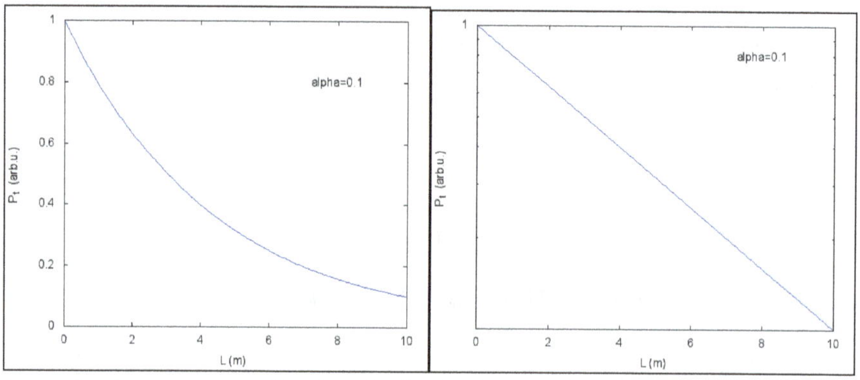

Transmitted power versus fiber length. (Left) Linear scale. (Right) Logarithmic scale.

Chromatic Dispersion

Chromatic dispersion originates from the interaction between electromagnetic waves (EMWs) and the bound electrons of the material that produces a frequency-dependent response of the medium as the refraction index is frequency dependent, i.e., $n = n(\omega)$. This occurs because of the resonance frequencies in the absorption spectrum. Far from the resonances the Sellmeier-formula can be used for n; this is an empirical relationship between the wavelength and the refractive index of a particular transparent medium:

$$n^2(\lambda) = 1 + \sum_{i=1}^{M} \frac{A_i\lambda^2}{\lambda^2 - \lambda_i^2}$$

where A_i and λ_i are the Sellmeier-coefficients. A_i represents the strength of the ith resonance while λ_i is the ith resonance wavelength. For $M = 3$ the formula is called Sellmeier-equation.

The Sellmeier-coefficients of a material can be determined experimentally. As an example, in figure the refractive index versus the wavelength is shown for glasses with different dopants.

Chromatic dispersion is especially critical in short-pulse propagation: a short pulse comprises many frequency components, therefore their speed is different according to $v = c / n(\omega)$. This phenomenon is present even without nonlinearity, but dispersion together with nonlinearity leads to completely different behavior of the fiber. As the frequency components of a pulse travel at different speed, one may wish to characterize this with one quantity:

$$v_g = \frac{1}{\beta_1} = \frac{c}{n_g} \equiv \frac{\partial \omega}{\partial k} = \frac{c}{n + \omega \dfrac{\partial n}{\partial \omega}}.$$

This is called *group velocity*, and it describes the speed of the envelope of an optical pulse. It can be derived from the Taylor-series of the mode propagation constant β around ω_0: the coefficient of the first term is $\beta_1 = d\beta/d\omega|_{\omega=\omega_0} = 1/v_g$.

In a transparent medium the group velocity of light depends on the optical wavelength or frequency; this phenomenon is called *group velocity dispersion* (GVD). Quantitatively it is the derivative of the inverse group velocity $1/v_g$ with respect to the angular frequency ω. It is the phenomenon that causes pulse broadening. The coefficient of the second term in the Taylor-series, $\beta_2 = d^2\beta/d\omega^2\big|_{\omega=\omega_0}$, is the parameter that describes GVD: it is called group velocity dispersion parameter.

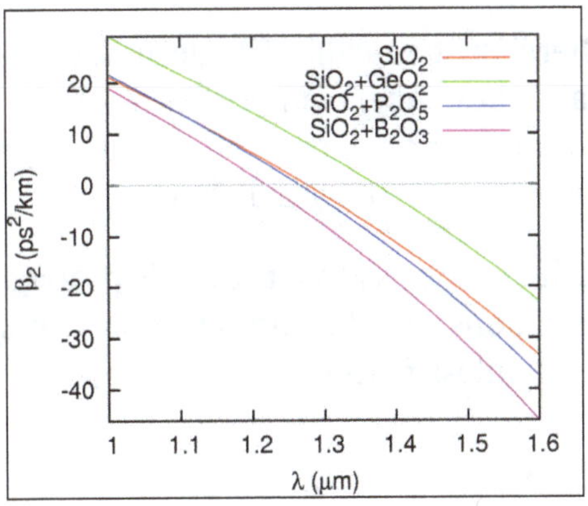

Group velocity dispersion parameter β_2 as a function of the wavelength λ for various fiber-optic materials.

By plotting β_2 as a function of the wavelength one can see that at a certain wavelength the value of β_2 changes sign. This wavelength is called zero-dispersion wavelength and denoted as λ_D. This value divides the normal and anomalous dispersion regime.

- Normal dispersion: When $\beta_2 > 0$, the dispersion is called normal dispersion. In this regime the blue-shifted components of an optical pulse travel slower.
- Anomalous dispersion: If $\beta_2 < 0$, anomalous dispersion occurs. Its effect is

opposite to the normal-dispersion regime, i.e., red-shifted components travel slower. In case of optical fibers solitons are supported in this regime keeping an equilibrium of dispersive and nonlinear effects.

There are many proposals to manipulate dispersion in fibers, and accordingly different types of fibers have been invented.

- Dispersion-shifted fibers (DSF): In these fibers the zero dispersion wavelength λ_D is shifted to the minimum fiber loss region in order to optimize both low dispersion and low loss.

- Dispersion-compensating fibers (DCF): In these fibers GVD is shifted to the $\lambda > 1.6\mu\text{m}$ wavelength region and exhibit $\beta_2 \gg 0$.

- Dispersion-flattened fibers (DFF): These usually multi-cladded fibers show low dispersion over a relatively wide wavelength range ($\lambda = 1.3 - 1.6\mu\text{m}$).

- Dispersion-decreasing fibers (DDF): In this case the D dispersion parameter defined as $d\beta_1/d\lambda$, and accordingly group velocity dispersion, varies (decreases) along the fiber length; this can be achieved by changing (decreasing) the core diameter with length.

Polarization-mode Dispersion

Earlier it was said that there exist so-called single-mode fibers. This name is not exact in the sense that actually these fibers support two orthogonal modes. Ideally these modes do not convert / couple energy to each other, but imperfections / anisotropy of real fibers result in mixing of these orthogonal states.

Modal birefringence is the fiber property that the mode propagation constant is different for the two polarization directions. It determines the circular symmetry of the fiber and is characterized by a dimensionless parameter:

$$B_m = \frac{|\beta_x - \beta_y|}{k_0} = |n_x - n_y|$$

where the polarization directions are assumed to be x and y; n_x and n_y are modal refractive indices. In accordance with the difference in the mode propagation constant, the speed of the modes are also different. This leads us to the definition of fast and slow axes: if $n_x < n_y$ then x is called the fast axis while y is the slow axis of the fiber.

In the case of short pulses the difference in the propagation speed in the two directions leads to pulse broadening which is known as polarization-mode dispersion (PMD). Naturally, estimation of the broadening is important and, assuming that

L is the fiber length, B_m is constant, and ΔT is the time delay between the two polarization components, this can be done by:

$$\Delta T = \left| \frac{L}{v_{g,x}} - \frac{L}{v_{g,y}} \right| = L \left| \beta_{1,x} - \beta_{1,y} \right| = L\,\Delta\beta_1.$$

In this formula B_m was assumed to be constant; in real fibers, however, this quantity changes randomly owing to the imperfections of the fiber which leads to the equalization of ΔT. Therefore variance can be used to measure the broadening: $\sigma_T^2 = \left\langle (\Delta T)^2 \right\rangle$. Through the variance a PMD parameter D_p can be introduced: assume that the correlation length of the fiber l_c is much smaller than the fiber length L, i.e., $l_c \ll L$, then:

$$\sigma_T \approx\sim D_p \sqrt{L},\ D_p = \Delta\beta_1 \sqrt{2l_c}$$

Usually the phenomenon of PMD is negligible; however, around λ_D for long fibers it becomes a phenomenon to be taken into account.

Finally we mention that there exist so-called polarization-maintaining fibers that preserve the polarization of the input light throughout transmission.

Fiber Nonlinearities

Consider an intense electromagnetic (EM) field falling to a dielectric; in that case non-linear dielectric response can be observed.

As the EM field causes anharmonic motion of bound electrons, the total polarization $\vec{P}$ can be written as:

$$\vec{P} = \epsilon_0\, \hat{\chi}(\vec{E})\vec{E}$$

$$\hat{\chi}(\vec{E}) = \hat{\chi}^{(1)} + \hat{\chi}^{(2)} \cdot \vec{E} + \hat{\chi}^{(3)} : \vec{E}\vec{E} + \ldots$$

where ϵ_0 is the vacuum permittivity and $\hat{\chi}^i$ are rank-$(i+1)$ tensors representing the i-th order susceptibility.

- $\hat{\chi}^{(1)}$ is called linear susceptibility; this quantity depends on the refractive index n and the attenuation coefficient α.
- $\hat{\chi}^{(2)}$ is the quadratic (or second-order) susceptibility; this part is present only in media that lack inversion symmetry in molecular level. Quadratic susceptibility is responsible for second harmonic generation and sum-frequency generation.

- Lowest order nonlinearities can be attributed to the effect of the third order susceptibility $\hat{\chi}^{(3)}$ leading to such phenomena as nonlinear refraction, third-harmonic generation (THG) or four-wave mixing (FWM).

Nonlinear Refraction

This phenomenon, which shows the intensity dependence of the refractive index, is caused by the third-order susceptibility. The effective refractive index of the fiber modes is:

$$n(\omega, I) = n(\omega) + n_2 I,$$

i.e., n has a linear part depending on the frequency and a nonlinear part proportional to the optical intensity I. The nonlinear-index coefficient n_2 is related to $\hat{\chi}^{(3)}$ through:

$$n_2(\omega; \Omega) = \frac{3}{8n_0}\mathrm{Re}(\hat{\chi}^{(3)})$$

where n_0 is the refractive index at the carrier frequency of an optical pulse.

The intensity dependence of n leads to two important phenomenon: self-phase modulation and cross-phase modulation.

- Self-phase modulation (SPM): As an ultrafast pulse travels in a medium, it induces a varying refractive index; in exchange, it produces a phase shift in the pulse. Its magnitude can be calculated as:

$$\phi = (n + n_2 I)k_0 L,$$

denoting the length of the fiber with L, the wave number with $k_0 = 2\pi/\lambda$, and the wave-length with λ. The formula shows that the SPM causes a nonlinear phase shift that is intensity dependent: $\phi_{\mathrm{SPM}} = n_2 |E|^2 k_0 L$. The effect of SPM can be observed through the spectral broadening of ultrashort pulses, or the formation of optical solitons.

- Cross-phase modulation (XPM): In order to observe XPM, two EM fields with different wavelength, direction or state of polarization are required. In this case one of these fields experiences nonlinear phase shift in the presence of the other. Assume that two optical fields of angular frequencies ω_1 and ω_2 travel through an optical fiber in the $\vec{x}$ direction; then the total electric field can be written as:

$$\vec{E} = \frac{1}{2}\vec{x}\sum_{k=1}^{2} E_k \exp(\pm i\omega_k t)$$

and the nonlinear phase shift originating from the XPM for the EM field ω_1 is:

$$\phi_{XPM} = 2n_2\, k_0 L |E_2|^2$$

Naturally, the total phase shift can be calculated as the sum of the SPM and XPM induced terms. XPM causes asymmetric spectral broadening of the co-propagating pulses, inter-channel crosstalk in WDM systems, and it can produce amplitude and timing jitter. More information on the topic concerning XPM in fibers can be found in.

Figure of Merit

The efficiency of the nonlinear processes in a fiber can be characterized by the figure of merit $I_0 L_{eff}$, where L_{eff} is the effective length of the interaction region and I_0 is the intensity. As the incident power can be expressed through I_0 and the spot area A as $P_0 = I_0 \pi A$, one can conclude that using tighter focus, i.e., smaller $A = \omega_0^2 \pi$, ω_0 being the spot radius, would lead to larger value of I_0; however, tighter focus also causes L_{eff} to be smaller.

For Gaussian beams that are widely used in optics to model laser radiation the effective length is the so-called Rayleigh range $z_R = \frac{w_0^2 \pi}{\lambda}$ that represents half the distance between the two coordinates along the propagation direction where the beam size increases to $\sqrt{2}\omega_0$, ω_0 being the waist size. In this case the figure of merit is:

$$I_0 L_{eff} = \frac{P_0}{\pi \omega_0^2} \frac{\omega_0^2 \pi}{\lambda} = \frac{P_0}{\lambda}$$

Owing to fiber losses, a different formula has to be used for describing fibers. The intensity is considered to be the function of distance z; assuming α to be the fiber loss one may express I as $I(z) = I_0 \exp(-\alpha z)$. Integrating $I(z)$ along the fiber we get the figure of merit of the fiber. Assuming long fibers, i.e., $\alpha L \gg 1$, the figure of merit is $IL_{eff} = \frac{P_0}{\alpha \pi \omega_0^2}$. This enables us to give a measure to nonlinear efficiency in the fiber:

$$\frac{(I\, L_{eff})}{(I_0\, L_{eff})} = \frac{\lambda}{\omega_0^2 \alpha \pi}.$$

The formula shows that reducing the waist size, i.e., the core radius of the fiber leads to higher nonlinearities that enables us to create so-called highly nonlinear fibers.

Pulse Propagation in Optical Fibers

As we mentioned previously, dispersion effects are much more intense for ultrashort pulses, therefore such pulses can be used for observing the nonlinear phenomena in fibers. In order to describe the propagation of ultrashort pulses in fibers one can use Maxwell's equations; in the slowly varying envelope approximation, and for pulses wider than 5 ps, these equations can be rearranged to get the following nonlinear Schrödinger equation (NSE):

$$\frac{i}{2}\beta_2\frac{\partial^2 A(z,t)}{\partial t^2}+\frac{\partial A(z,t)}{\partial z}=-\frac{\alpha}{2}A(z,t)+i\gamma\left|A(z,t)\right|^2 A(z,t),$$

where β_2 is the GVD parameter, $A(z,t)$ denotes the slowly varying envelope of the pulse, α is the loss parameter, while γ is the nonlinear parameter:

$$\gamma=\frac{\omega n_2}{A_{\text{eff}}c},$$

A_{eff} being the effective area of the fiber core. As the dispersive and nonlinear phenomena are much more intense for shorter pulses, the above NSE can be used only by taking into account further terms responsible for higher order nonlinearity.

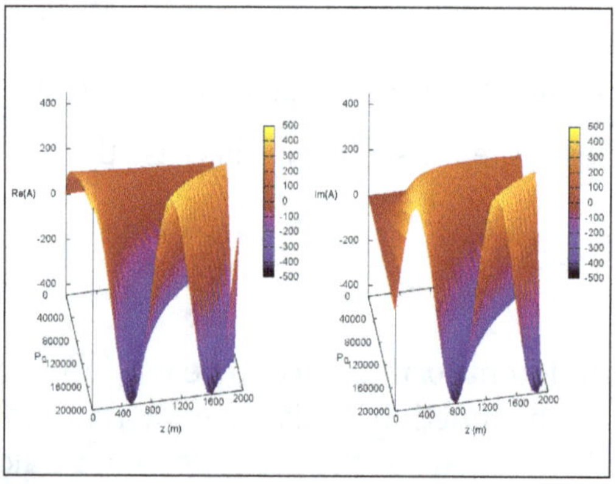

The time-independent result A of the NSE as a function of the power P_0 and the distance Z.

For continuous waves (CW) of power P_0 and $\alpha=0$, i.e., without losses, the NSE of the fiber can be solved easily. Substituting the SPM induced nonlinear phase shift $\varphi_{\text{NL}}=P_0 z\gamma$, the time-independent result is:

$$A(z)=\sqrt{P_0}\exp(i\gamma P_0 z).$$

The stability of the CW solution strongly depends on the type of GVD experienced by the wave inside the fiber. This can be shown by introducing a small perturbation a_p to the system; this appears in the right hand side of the NSE as $A(z,t)\rightarrow a+a_p$, while

$|A(z,t)|^2 \to P_0$. The general solution is a complex harmonic function containing the perturbation frequency Ω and the corresponding wave number K. After some calculation the dispersion relation for this case can be expressed as:

$$K(\Omega) = \pm \frac{1}{2}|\beta_2||\Omega|\sqrt{\frac{4P_0\gamma}{\beta_2} + \Omega^2}.$$

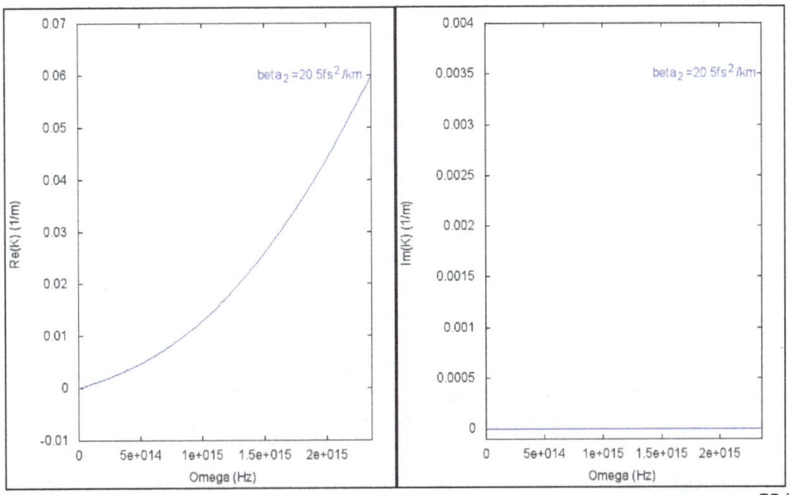

The dispersion relation (Left) Real part of $K(\Omega)$. (Right) Imaginary part of $K(\Omega)$.

Because of the square root in the above relation one can suspect that there may be parameters for which the value K is imaginary. Indeed, for $\beta_2 < 0$ and for appropriate values of Ω the value of K can be imaginary that leads to higher perturbations, i.e., less stable solutions. Such an instability is called modulation instability because depending on the power it causes the modulation of the beam.

Soliton Formation in Fibers

The modulation instability occurring in the anomalous GVD regime of optical fibers leads to solutions of the NSE which are called solitons. If we apply the inverse scattering method to the NSE, an exact solution can be found. Introducing the dispersion distance $L_D = T_0^2/|\beta_2|$ and the pulse width T_0 and by using soliton-units, i.e., $\tau = t/T_0$, $\xi = z/L_D$, and $u = A\sqrt{\gamma L_D}$, the NSE can be written in a normalized form. Without losses the canonical form of the NSE (CNSE) can be obtained:

$$i\frac{\partial u}{\partial \xi} \pm \frac{1}{2}\frac{\partial^2 u}{\partial \tau^2} + |u|^2 u = 0.$$

This is the equation that describes solitons; and according to the sign of the second-order term one may distinguish two types of solitons.

- Bright solitons (sign +). In this case the pulse propagating in the fiber experiences anomalous dispersion. For a sech-type input pulse, i.e., $u(0,\tau) = N\sec h(\tau)$, where N is the order of the soliton, the pulse propagating in the fiber is either unchanged ($N = 1$; fundamental solitons) or shows periodic oscillations of its temporal and spectral shape ($N > 1, N \in \mathbb{N}$; higher-order solitons). In the latter case time-dependent phase shift, i.e., frequency chirp can be observed; an example to this behavior is shown on the figure below.

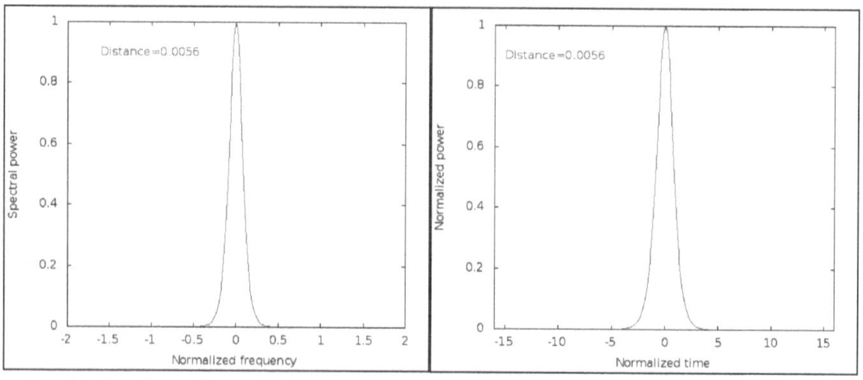

Evolution of a third order soliton over one soliton period. Left: Spectral power versus normalized frequency. Right: Normalized power versus normalized time. No chirp in the input pulse was assumed.

The evolution period of the higher-order solitons is called soliton period. Throughout the propagation the fundamental mode, except for a $\xi/2$ phase shift, remains unchanged, no chirp can be observed. Solitons are remarkably resistant to perturbations. This means that even an initially imperfect soliton can propagate so that the output pulse is close to ideal.

- Dark solitons (sign -): It occurs in the case of normal dispersion. The name dark soliton originates from the intensity profile where an unchanging dark spot can be observed throughout the propagation. The solution of the CNSE in that case is:

$$u_d(\xi,\tau) = \{u_0 \cos\phi \tanh[u_0 \cos\phi(\tau - u_0\xi \sin\phi] - iu_0 \sin\phi\} \exp(iu_0^2\xi),$$

where $\phi \in [0; \pi/2]$ is an internal phase angle, and u_0 is the background amplitude of the CW. The different values of ϕ define different types of dark solitons: the $\phi = 0$ case means that the power of the soliton in the middle of the dark spot is zero, i.e., a black spot can be observed, hence their name is black soliton; in contrast, for $\phi \neq 0$, the center of the spot is gray as the intensity is not zero, therefore these solitons are called gray solitons. As the phase changes along the width of dark solitons, these solutions are generally chirped.

Controlling the Dispersion

As dispersion is a basic phenomenon in fiber optics, its control is very important. If a fiber is created such that the GVD varies along its length, the average GVD can be reduced which leads to the negligibility of higher-order dispersion phenomena. Such a fiber can be produced by joining fiber segments of constant dispersion. Although the GVD of a particular section may be high, the average can be sufficiently low. This technique is called dispersion management.

The NSE of the solitons describing such a fiber must contain terms corresponding to the GVD change along the fiber and the fiber losses which leads to:

$$\frac{1}{2}p(\xi)\frac{\partial^2 u}{\partial \tau^2}+i\frac{\partial u}{\partial \xi}=-\frac{i}{2}\Gamma u-|u|^2 u.$$

In the above equation ξ is a normalized distance, $L_D = T_0^2/|\beta_2(0)|$ is the dispersion length used for normalizing ξ, while $p(\xi)=|\beta_2(\xi)/\beta_2(0)|$ is the GVD at ξ. Denoting the definite integral of $p(x)$ over the interval $x=[0,\xi]$ by ξ' and substituting $u=v\sqrt{p}$ in the above equation, i.e., renormalizing amplitude and length to the local GVD, the resulting equation is:

$$\frac{1}{2}\frac{\partial^2 v}{\partial \tau^2}+\frac{\partial v}{\partial \xi'}=\frac{iv}{2p}\left(\Gamma+\frac{dp}{d\xi'}\right)-|v|^2 v.$$

Using properly chosen parameters all the losses of the fiber can be eliminated, therefore soliton formation and propagation is possible even in lossy fibers. The criterion for maintaining the balance between GVD and SPM is $\frac{1}{2}\frac{\partial^2 v}{\partial \tau^2}+\frac{\partial v}{\partial \xi'}=\frac{iv}{2p}\left(\Gamma+\frac{dp}{d\xi'}\right)-|v|^2 v.$, z being the distance along the fiber; these fibers are called dispersion-decreasing fibers.

Material and Fabrication

Now that we got familiar with the main problems and phenomena occurring in fibers, we can concentrate on how to create such fibers, or what techniques do we have at all to fabricate fibers.

As fused silica glass shows characteristics with considerably low loss, this is the primary material in fiber technique. Naturally, somehow we have to achieve the desired refractive index difference of the core and the cladding (or a special refractive index profile); to this end dopants may be used. Germanium dioxide and phosphorus pentoxide (P_2O_5) causes the refractive index to increase, therefore these materials are suitable for producing the core; on the other hand, using boron (B) or fluorine (F) as dopants the refractive index decreases which is needed for the cladding. In case of fiber amplifiers or lasers rare-earth ions such as erbium trichloride $(ErCl_3)$ or neodymium sesquioxide (Nd_2O_3) are used.

The method of fabrication usually consists of two steps:

- Creating a preform: A cylindrical preform with the appropriate refractive index profile but (compared to the final dimensions of the fiber) larger radial and smaller axial size is created. The length of the preform is around 1 m while its diameter is in the cm range. The desired refractive index distribution can be created by vapor deposition. There are various vapor deposition methods such as modified chemical vapor deposition (MCVD or just CVD), outside vapor deposition (OVD), vapor phase axial deposition (VAD or AVD) or plasma chemical vapor deposition (PCVD). These methods differ in various aspects such as the possible purity of the material, the refractive index control, (GeO_2) deposition efficiency or the mechanical strength of the fiber. Vapor deposition methods guarantee extremely low propagation losses because the purity of the materials can be very high. Note that there are materials for which the use of vapor deposition techniques is not possible; in this case the rod-in-tube technique can be used.

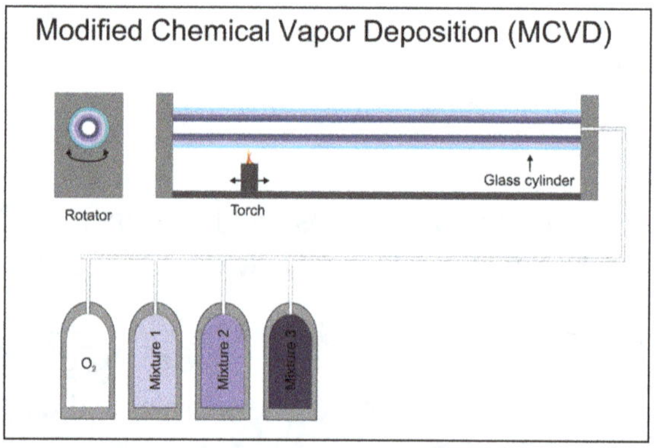

Schematic diagram of the modified chemical vapor deposition (MCVD) technique.

- Drawing: When the preform is ready, that is, we have a cylinder with proper radial refractive index distribution corresponding to the radial distribution of the fiber to be created, the object can be pulled in an apparatus of several meters high called fiber-drawing tower. At the top of the fiber-drawing tower a furnace (or oven) is located. Throughout the process the preform is heated close to its melting point, which enables us to pull a thin and very long (of length in the km range) fiber out of the lower part of the preform. During the process the pulling speed and the heating temperature is adjusted automatically, thus the fiber diameter remains constant.

As a final step, the fiber is “got dressed”, that is, a polymer coating is added for mechanical and chemical protection. Acrylate, silicone or polyimide (PI) is used as coating material.

Note that there are fabrication methods in which no preform is used such as the double crucible method; in this case the materials for the core and the cladding are stored in separate reservoirs and are simultaneously drawn from the crucible. It is more adaptable to different materials than the vapor deposition method; however, the crucible material may contaminate the fiber material therefore producing low-loss ultra-pure fibers with this method is a challenging problem.

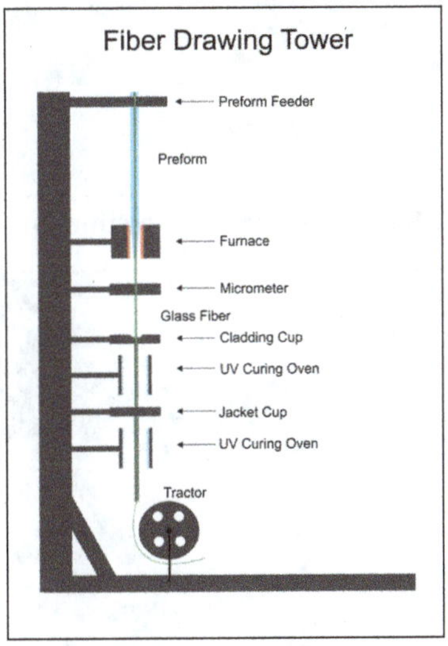

Diagram of a fiber drawing tower.

Thin-film Optics

Thin-film optics is the branch of optics that deals with very thin structured layers of different materials. In order to exhibit thin-film optics, the thickness of the layers of material must be on the order of the wavelengths of visible light (about 500 nm). Layers at this scale can have remarkable reflective properties due to light wave interference and the difference in refractive index between the layers, the air, and the substrate. These effects alter the way the optic reflects and transmits light. This effect, known as thin-film interference, is observable in soap bubbles and oil slicks.

More general periodic structures, not limited to planar layers, are known as photonic crystals.

In manufacturing, thin film layers can be achieved through the deposition of one or more thin layers of material onto a substrate (usually glass). This is most often done using a physical vapor deposition process, such as evaporation or sputter deposition, or a chemical process such as chemical vapor deposition.

Thin films are used to create optical coatings. Examples include low emissivity panes of glass for houses and cars, anti-reflective coatings on glasses, reflective baffles on car headlights, and for high precision optical filters and mirrors. Another application of these coatings is spatial filtering.

Dichroic filters are created using thin film optics.

Thin film interference caused by ITO defrosting coating on an Airbus cockpit window. The film thickness is intentionally non-uniform to provide even heating at different distances from the electrodes.

A pattern of coloured light formed by interference between white light being reflected from the surface of a thin film of diesel fuel on the surface of water, and the diesel-water interface.

Hafnium oxidized ingots which exhibits thin film optical effects.

Examples in the Natural World

The blue wing patches of the Aglais io.

Graphium sarpedon.

The breast feathers of the Lawes's parotia.

The thin-film interference that can be seen on many insect wings is due to thin-film optics.

The glossy flowers of Ranunculus buttercups.

Thin-film layers are common in the natural world. Their effects produce colors seen in soap bubbles and oil slicks, as well as the structural coloration of some animals. The wings of many insects act as thin-films, because of their minimal thickness. This is clearly visible in the wings of many flies and wasps. In butterflies, the thin-film optics is visible when wing itself is not covered by wing scales, which is the case in the blue wing spots of the Aglais io and the blue-green patches of the Graphium sarpedon. In buttercups, the flower's gloss is due to a thin-film, which enhances the flower's visibility to pollinating insects and aids in temperature regulation of the plant's reproductive organs.

Crystal Optics

Crystal optics is the study of the propagation of light, and associated phenomena, in crystalline solids. For a simple cubic crystal the atomic arrangement is such that in each direction through the crystal the crystal presents the same optical appearance. The atoms in anisotropic crystals are closer together in some planes through the material than in others. In anisotropic crystals the optical characteristics are different in different directions. In classical physics the progress of an electromagnetic wave through a material involves the periodic displacement of electrons. In anisotropic substances the forces resisting these displacements depend on the displacement direction. Thus the velocity of a light wave is different in different directions and for different states of polarization. The absorption of the wave may also be different in different directions.

In an isotropic medium the light from a point source spreads out in a spherical shell. The light from a point source embedded in an anisotropic crystal spreads out in two wave surfaces, one of which travels at a faster rate than the other. The polarization of the light varies from point to point over each wave surface, and in any particular direction from the source the polarization of the two surfaces is opposite. The characteristics of these surfaces can be determined experimentally by making measurements on a given crystal.

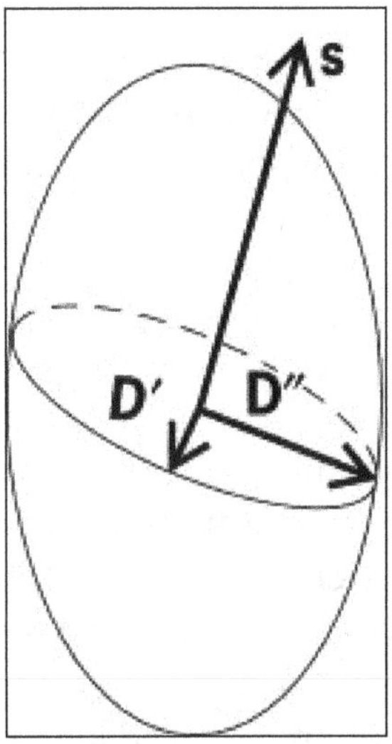

Index ellipsoid, showing construction of directions of vibrations of D vectors belonging to a wave normal s.

In the most general case of a transparent anisotropic medium, the dielectric constant is different along each of three orthogonal axes. This means that when the light vector is oriented along each direction, the velocity of light is different. One method for calculating the behavior of a transparent anisotropic material is through the use of the index ellipsoid, also called the reciprocal ellipsoid, optical indicatrix, or ellipsoid of wave normals. This is the surface obtained by plotting the value of the refractive index in each principal direction for a linearly polarized light vector lying in that direction. The different indices of refraction, or wave velocities associated with a given propagation direction, are then given by sections through the origin of the coordinates in which the index ellipsoid is drawn. These sections are ellipses, and the major and minor axes of the ellipse represent the fast and slow axes for light proceeding along the normal to the plane of the ellipse. The length of the axes represents the refractive indices for the fast and slow wave, respectively. The most asymmetric type of ellipsoid has three unequal axes. It is a general rule in crystallography that no property of a crystal will have less symmetry than the class in which the crystal belongs.

Accordingly, there are many crystals which, for example, have four- or sixfold rotation symmetry about an axis, and for these the index ellipsoid cannot have three unequal axes but is an ellipsoid of revolution. In such a crystal, light will be propagated along this axis as though the crystal were isotropic, and the velocity of propagation will be independent of the state of polarization. The section of the index ellipsoid at right angles to this direction is a circle. Such crystals are called uniaxial and the mathematics of their optical behavior is relatively straightforward.

In crystals of low symmetry the index ellipsoid has three unequal axes. These crystals are termed biaxial and have two directions along which the wave velocity is independent of the polarization direction. These correspond to the two sections of the ellipsoid which are circular.

The normal to a plane wavefront moves with the phase velocity. The Huygens wavelet, which is the light moving out from a point disturbance, will propagate with a ray velocity. Just as the index ellipsoid can be used to compute the phase or wave velocity, so can a ray ellipsoid be used to calculate the ray velocity. The length of the axes of this ellipsoid is given by the velocity of the linearly polarized ray whose electric vector lies in the axis direction.

The refraction of a light ray on passing through the surface of an anisotropic uniaxial crystal can be calculated with Huygens wavelets in the same manner as in an isotropic material. For the ellipsoidal wavelet this results in an optical behavior which is completely different from that normally associated with refraction. The ray associated with this behavior is termed the extraordinary ray. At a crystal surface where the optic axis is inclined at an angle, a ray of unpolarized light incident normally on the surface is split into two beams: the ordinary ray, which proceeds through the surface without deviation; and the extraordinary ray, which is deviated by an angle determined by a line drawn from the center of one of the Huygens ellipsoidal wavelets to the point at which the ellipsoid is tangent to a line parallel to the surface. The two beams are oppositely linearly polarized.

Quantum Optics

Quantum optics is a field of quantum physics that deals specifically with the interaction of photons with matter. The study of individual photons is crucial to understanding the behavior of electromagnetic waves as a whole.

To clarify exactly what this means, the word "quantum" refers to the smallest amount of any physical entity that can interact with another entity. Quantum physics, therefore, deals with the smallest particles; these are incredibly tiny sub-atomic particles which behave in unique ways.

The word "optics," in physics, refers to the study of light. Photons are the smallest particles of light (though it is important to know that photons can behave as both particles and waves).

Development of Quantum Optics and the Photon Theory of Light

The theory that light moved in discrete bundles (i.e. photons) was presented in Max Planck's 1900 paper on the ultraviolet catastrophe in black body radiation. In 1905, Einstein expanded on these principles in his explanation of the photoelectric effect to define the photon theory of light.

Quantum physics developed through the first half of the twentieth century largely through work on our understanding of how photons and matter interact and inter-relate. This was viewed, however, as a study of the matter involved more than the light involved.

In 1953, the maser was developed (which emitted coherent microwaves) and in 1960 the laser (which emitted coherent light). As the property of the light involved in these devices became more important, quantum optics began being used as the term for this specialized field of study.

Findings

Quantum optics (and quantum physics as a whole) views electromagnetic radiation as traveling in the form of both a wave and a particle at the same time. This phenomenon is called wave-particle duality. The most common explanation of how this works is that the photons move in a stream of particles, but the overall behavior of those particles is determined by a quantum wave function that determines the probability of the particles being in a given location at a given time.

Taking findings from quantum electrodynamics (QED), it is also possible to interpret quantum optics in the form of the creation and annihilation of photons, described by field operators. This approach allows the use of certain statistical approaches that are useful in analyzing the behavior of light, although whether it represents what is physically taking place is a matter of some debate (although most people view it as just a useful mathematical model).

Applications

Lasers (and masers) are the most obvious application of quantum optics. Light emitted from these devices is in a coherent state, which means the light closely resembles a classical sinusoidal wave. In this coherent state, the quantum mechanical wave function (and thus the quantum mechanical uncertainty) is distributed equally. The light emitted from a laser is, therefore, highly ordered, and generally limited to essentially the same energy state (and thus the same frequency & wavelength).

References

- Optics, physics: byjus.com, Retrieved 21 June, 2019
- Van der Kooi, C.J.; Elzenga, J.T.M.; Dijksterhuis, J.; Stavenga, D.G. (2017). "Functional optics of glossy buttercup flowers". Journal of the Royal Society Interface. 14 (127): 20160933. Doi:10.1098/rsif.2016.0933. PMC 5332578. PMID 28228540.
- Introduction, FJ-Nonlinear-Optics, TAMOP: physics.ttk.pte.hu, Retrieved 22 July, 2019
- 4-secondorder-interactions, FJ-Nonlinear-Optics, TAMOP: physics.ttk.pte.hu, Retrieved 23 August, 2019
- 9-nonlinear-optics-with-pulses, FJ-Nonlinear-Optics, TAMOP, physics.ttk.pte.hu, Retrieved 24 January, 2019
- Crystal+Optics: thefreedictionary.com, Retrieved 25 February, 2019
- What-is-quantum-optics-2699361: thoughtco.com, Retrieved 26 March, 2019

2
Geometrical Optics

Geometrical optics is a model of optics that describes light propagation in terms of rays. Laws of reflection, refraction, total internal reflection, etc. are studied under geometrical optics. This chapter has been carefully written to provide an easy understanding of geometrical optics.

Light can behave in many ways like a wave or a packet of photons which travel in a straight line exhibiting the property of rectilinear propagation. Geometrical optics deals with the propagation of light in a straight line and phenomena such as reflection, refraction, polarization, etc. A ray of light gives the direction of propagation of light. In the absence of an obstacle, the rays advance in a straight line without changing direction. When light meets a surface separating two transparent media, reflection and refraction occur and the light rays bend.

Reflection at Smooth Surfaces

A light ray is reflected by a smooth surface in accordance with the 2 laws of reflection:

- The angle of incidence is always equal to the angle of reflection.
- The incident ray, the reflected ray, and the normal to the reflecting surface are coplanar. The images formed by a plane mirror are equidistant from the distance between the object and the mirror.

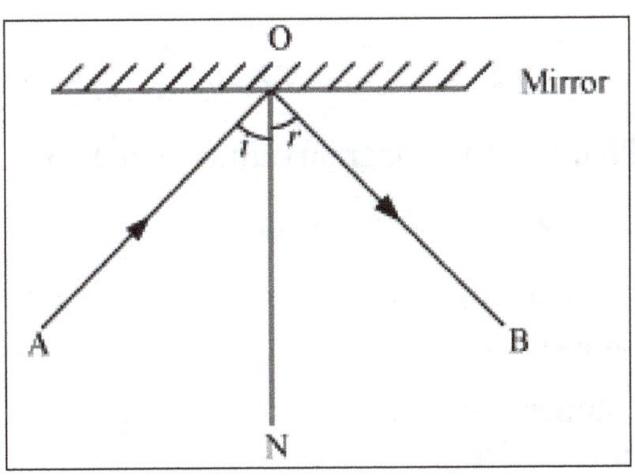

The laws of reflection are the same for plane and curved surfaces. A normal can be drawn from any point of the curved surface by first drawing the tangent plane from that point and then drawing the line perpendicular to that plane. Angles of incidence and reflection are defined from this normal. The angle of incidence is equal to angle of reflection.

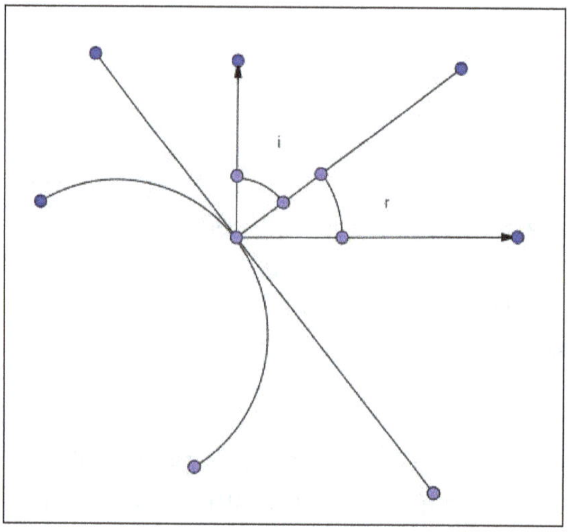

Spherical Mirrors

A spherical mirror is a part cut from a hollow sphere. The spherical mirrors are generally constructed from glass. One surface of the glass is silvered. If reflection takes place at the convex surface, it is called a convex mirror and if it takes place on the concave surface, it is called a concave mirror.

Terms used in Spherical Mirrors

- Center of curvature (C): It is the center of the sphere from which the mirror is made.
- Focus (f): It is the point at which the rays meet (either virtually or physically). Although we have not verified, from further exploration we will understand that $R=2f$.
- Pole (P): It is the surface of the mirror on which light falls.
- Principle axis (PA): It is the imaginary line which is perpendicular to the pole.
- Paraxial rays: These are rays parallel and close to the principal axis.
- u: This is the distance between the object and the pole and is sometimes referred to as the object distance.
- v: This is the distance between the pole and the image and is sometimes referred to as the image distance.

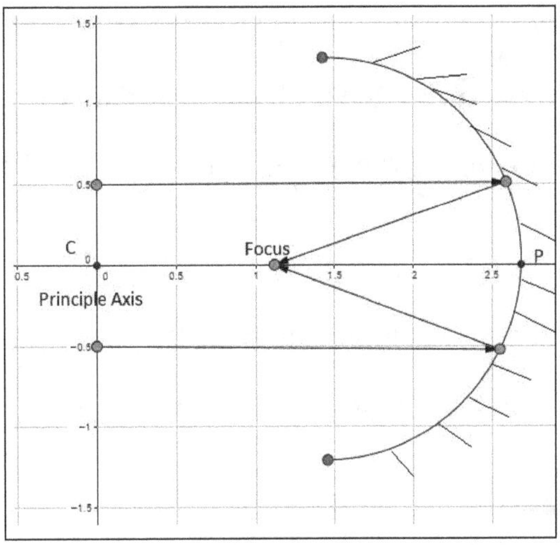

Here (0,0) is the C, and the principal axis is the xx-axis.

Sign Conventions

Cartesian Sign Conventions

- The pole is taken to be the origin.
- The direction in which the light moves is taken as the positive direction.
- The height above the principal axis is taken as positive while the height below it is taken as negative.

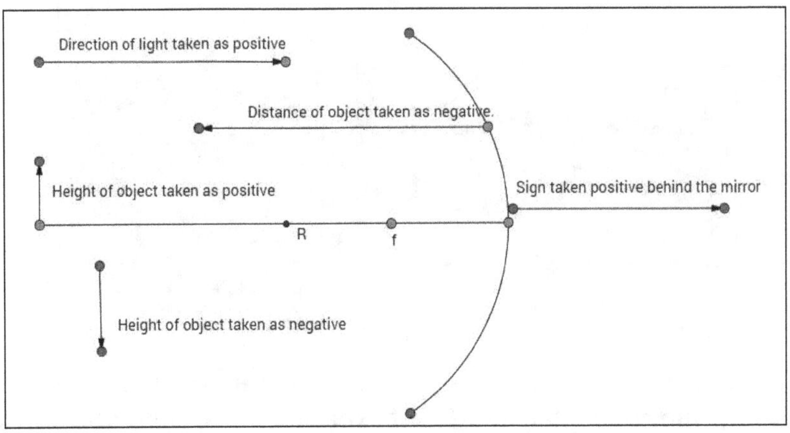

Mirror Formula

The main formula used for solving problems on mirrors is:

$$\frac{1}{v}+\frac{1}{u}=\frac{1}{f}.$$

While solving problems, do not forget to apply sign conventions and make sure you always take the object distance, u, as negative only. See the examples below which will make you comfortable with solving the problems based on the mirror formula.

The magnification of a mirror is given by $\frac{h_i}{h_o}$, where h_i is height of the image and h_o is height of the object.

Magnification is also given by $-\frac{v}{u}$.

Again we should remember not to forget applying the sign conventions.

If magnification is *-ve*, it means the formed image is inverted.

Reflection of Light

Reflection of light (and other forms of electromagnetic radiation) occurs when the waves encounter a surface or other boundary that does not absorb the energy of the radiation and bounces the waves away from the surface. The simplest example of visible light reflection is the surface of a smooth pool of water, where incident light is reflected in an orderly manner to produce a clear image of the scenery surrounding the pool. Throw a rock into the pool, and the water is perturbed to form waves, which disrupt the reflection by scattering the reflected light rays in all directions.

Reflections From the Surface of Water

Smooth Water Surface — Wavy Water Surface

Some of the earliest accounts of light reflection originate from the ancient Greek mathematician Euclid, who conducted a series of experiments around 300 BC, and appears to have had a good understanding of how light is reflected. However, it wasn't until a millennium and a half later that the Arab scientist Alhazen proposed a law describing exactly what happens to a light ray when it strikes a smooth surface and then bounces off into space.

The incoming light wave is referred to as an incident wave, and the wave that is bounced away from the surface is termed the reflected wave. Visible white light that is directed onto the surface of a mirror at an angle (incident) is reflected back into space by the mirror surface at another angle (reflected) that is equal to the incident angle, as presented for the action of a beam of light from a flashlight on a smooth, flat mirror in figure. Thus, the angle of incidence is equal to the angle of reflection for visible light as well as for all other wavelengths of the electromagnetic radiation spectrum. This concept is often termed the Law of Reflection. It is important to note that the light is not separated into its component colors because it is not being "bent" or refracted, and all wavelengths are being reflected at equal angles. The best surfaces for reflecting light are very smooth, such as a glass mirror or polished metal, although almost all surfaces will reflect light to some degree.

Because light behaves in some ways as a wave and in other ways as if it were composed of particles, several independent theories of light reflection have emerged. According to wave-based theories, the light waves spread out from the source in all directions, and upon striking a mirror, are reflected at an angle determined by the angle at which the light arrives. The reflection process inverts each wave back-to-front, which is why a reverse image is observed. The shape of light waves depends upon the size of the light source and how far the waves have traveled to reach the mirror. Wavefronts that originate from a source near the mirror will be highly curved, while those emitted by distant light sources will be almost linear, a factor that will affect the angle of reflection.

According to particle theory, which differs in some important details from the wave concept, light arrives at the mirror in the form of a stream of tiny particles, termed photons, which bounce away from the surface upon impact. Because the particles are so small, they travel very close together (virtually side by side) and bounce from different points, so their order is reversed by the reflection process, producing a mirror image. Regardless of whether light is acting as particles or waves, however, the result of reflection is the same. The reflected light produces a mirror image.

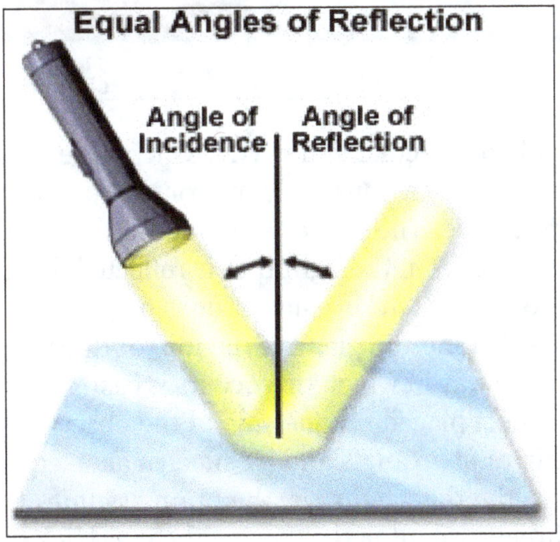

The amount of light reflected by an object, and how it is reflected, is highly dependent upon the degree of smoothness or texture of the surface. When surface imperfections are smaller than the wavelength of the incident light (as in the case of a mirror), virtually all of the light is reflected equally. However, in the real world most objects have convoluted surfaces that exhibit a diffuse reflection, with the incident light being reflected in all directions. Many of the objects that we casually view every day (people, cars, houses, animals, trees, etc.) do not themselves emit visible light but reflect incident natural sunlight and artificial light. For instance, an apple appears a shiny red color because it has a relatively smooth surface that reflects red light and absorbs other non-red (such as green, blue, and yellow) wavelengths of light. The reflection of light can be roughly categorized into two types of reflection. Specular reflection is defined as light reflected from a smooth surface at a definite angle, whereas diffuse reflection is produced by rough surfaces that tend to reflect light in all directions. There are far more occurrences of diffuse reflection than specular reflection in our everyday environment.

To visualize the differences between specular and diffuse reflection, consider two very different surfaces: a smooth mirror and a rough reddish surface. The mirror reflects all of the components of white light (such as red, green, and blue wavelengths) almost equally and the reflected specular light follows a trajectory having the same angle from the normal as the incident light. The rough reddish surface, however, does not reflect all wavelengths because it absorbs most of the blue and green components, and reflects the red light. Also, the diffuse light that is reflected from the rough surface is scattered in all directions.

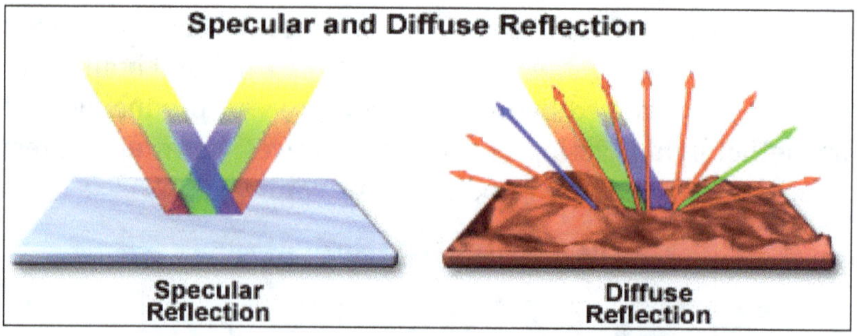

Perhaps the best example of specular reflection, which we encounter on a daily basis, is the mirror image produced by a household mirror that people might use many times a day to view their appearance. The mirror's smooth reflective glass surface renders a virtual image of the observer from the light that is reflected directly back into the eyes. This image is referred to as "virtual" because it does not actually exist (no light is produced) and appears to be behind the plane of the mirror due to an assumption that the brain naturally makes. The way in which this occurs is easiest to visualize when looking at the reflection of an object placed on one side of the observer, so that the light from the object strikes the mirror at an angle and is reflected at an equal angle to the viewer's eyes. As the eyes receive the reflected rays, the brain assumes that the light rays have reached the eyes in a direct straight path. Tracing the rays backward toward the mirror, the brain perceives an image that is positioned behind the mirror. An interesting feature of this reflection artifact is that the image of an object being observed appears to be the same distance behind the plane of the mirror as the actual object is in front of the mirror.

The type of reflection that is seen in a mirror depends upon the mirror's shape and, in some cases, how far away from the mirror the object being reflected is positioned. Mirrors are not always flat and can be produced in a variety of configurations that provide interesting and useful reflection characteristics. Concave mirrors, commonly found in the largest optical telescopes, are used to collect the faint light emitted from very distant stars. The curved surface concentrates parallel rays from a great distance into a single point for enhanced intensity. This mirror design is also commonly found in shaving or cosmetic mirrors where the reflected light produces a magnified image of the face. The inside of a shiny spoon is a common example of a concave mirror surface, and can be used to demonstrate some properties of this mirror type. If the inside of the spoon is held close to the eye, a magnified upright view of the eye will be seen (in this case the eye is closer than the focal point of the mirror). If the spoon is moved farther away, a demagnified upside-down view of the whole face will be seen. Here the image is inverted because it is formed after the reflected rays have crossed the focal point of the mirror surface.

Another common mirror having a curved-surface, the convex mirror, is often used in automobile rear-view reflector applications where the outward mirror curvature produces a smaller, more panoramic view of events occurring behind the vehicle. When parallel rays strike the surface of a convex mirror, the light waves are reflected outward so that they diverge. When the brain retraces the rays, they appear to come from behind the mirror where they would converge, producing a smaller upright image (the image is upright since the virtual image is formed before the rays have crossed the focal point). Convex mirrors are also used as wide-angle mirrors in hallways and businesses for security and safety. The most amusing applications for curved mirrors are the novelty mirrors found at state fairs, carnivals, and fun houses. These mirrors often incorporate a mixture of concave and convex surfaces, or surfaces that gently change curvature, to produce bizarre, distorted reflections when people observe themselves.

Spoons can be employed to simulate convex and concave mirrors, as illustrated in Figure for the reflection of a young woman standing beside a wooden fence. When the image of the woman and fence are reflected from the outside bowl surface (convex) of the spoon, the image is upright, but distorted at the edges where the spoon curvature varies. In contrast, when the reverse side of the spoon (the inside bowl, or concave, surface) is utilized to reflect the scene, the image of the woman and fence are inverted.

The reflection patterns obtained from both concave and convex mirrors are presented in Figure. The concave mirror has a reflection surface that curves inward, resembling a portion of the interior of a sphere. When light rays that are parallel to the principal or optical axis reflect from the surface of a concave mirror (in this case, light rays from the owl's feet), they converge on the focal point (red dot) in front of the mirror. The distance from the reflecting surface to the focal point is known as the mirror's focal length. The size of the image depends upon the distance of the object from the mirror and its position with respect to the mirror surface. In this case, the owl is placed away from the center of curvature and the reflected image is upside down and positioned between the mirror's center of curvature and its focal point.

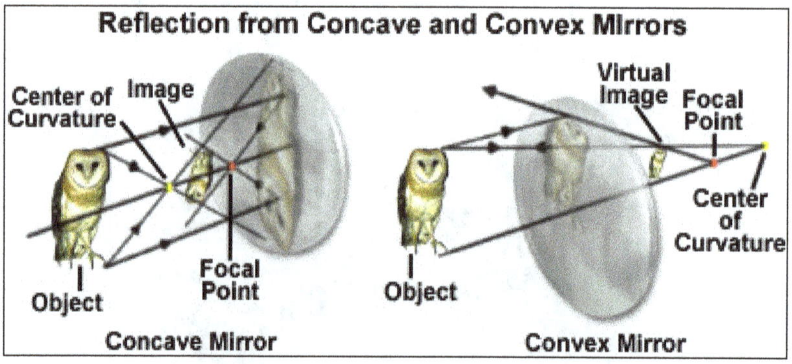

The convex mirror has a reflecting surface that curves outward, resembling a portion of the exterior of a sphere. Light rays parallel to the optical axis are reflected from the surface in a direction that diverges from the focal point, which is behind the mirror. Images formed with convex mirrors are always right side up and reduced in size. These images are also termed virtual images, because they occur where reflected rays appear to diverge from a focal point behind the mirror.

The manner in which gemstones are cut is one of the more aesthetically important and pleasing applications of the principles of light reflection. Particularly in the case of diamonds, the beauty and economic value of an individual stone is largely determined by the geometric relationships of the external faces (or facets) of the gem. The facets that are cut into a diamond are planned so that most of the light that falls on the front face of the stone is reflected back toward the observer . A portion of the light is reflected directly from the outside upper facets, but some enters the diamond, and after internal reflection, is reflected back out of the stone from the inside surfaces of the lower facets. These internal ray paths and multiple reflections are responsible for a diamond's sparkle, often referred to as its "fire". An interesting consequence of a perfectly cut stone is that it will show a brilliant reflection when viewed from the front, but will look darker or dull from the back.

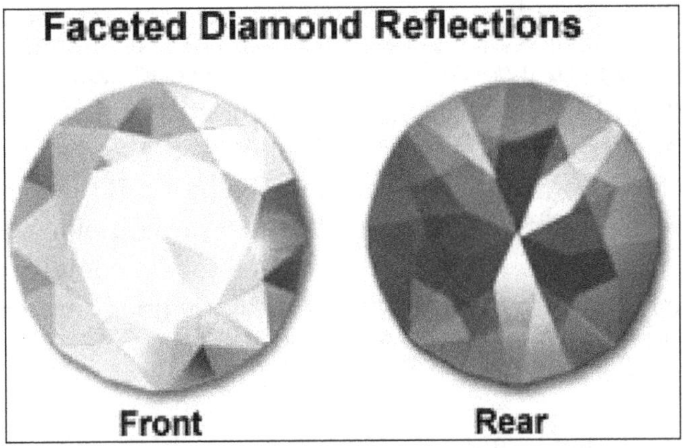

Light rays are reflected from mirrors at all angles from which they arrive. In certain other situations, however, light may only be reflected from some angles and not others, leading to a phenomenon known as total internal reflection. This can be illustrated by a situation in which a diver working below the surface of perfectly calm water shines a bright flashlight directly upward at the surface. If the light strikes the surface at right angles it continues directly out of the water as a vertical beam projected into the air. If the light's beam is directed at a slight angle to the surface, so that it impacts the surface at an oblique angle, the beam will emerge from the water, but will be bent by refraction toward the plane of the surface. The angle between the emerging beam and the surface of the water will be smaller than the angle between the light beam and the surface below the water.

If the diver continues to angle the light at more of a glancing angle to the surface, the beam rising out of the water will get closer and closer to the surface, until at some point it will be parallel to the surface. Because of light bending due to refraction, the emerging beam will become parallel to the surface before the light below the water has reached the same angle. The point at which the emerging beam becomes parallel to the surface occurs at the critical angle for water. If the light is angled still further, none of it will emerge. Instead of being refracted, all of the light will reflect at the water's surface back into the water just as it would at the surface of a mirror.

The principle of total internal reflection is the basis for fiber optic light transmission that makes possible medical procedures such as endoscopy, telephone voice transmissions encoded as light pulses, and devices such as fiber optic illuminators that are widely used in microscopy and other tasks requiring precision lighting effects. The prisms employed in binoculars and in single-lens reflex cameras also utilize total internal reflection to direct images through several 90-degree angles and into the user's eye. In the case of fiber optic transmission, light entering one end of the fiber is reflected internally numerous times from the wall of the fiber as it zigzags toward the other end, with none of the light escaping through the thin fiber walls. This method of "piping" light can be maintained for long distances and with numerous turns along the path of the fiber.

Total internal reflection is only possible under certain conditions. The light is required to travel in a medium that has relatively high refractive index, and this value must be higher than that of the surrounding medium. Water, glass, and many plastics are therefore suitable for use when they are surrounded by air. If the materials are chosen appropriately, reflections of the light inside the fiber or light pipe will occur at a shallow angle to the inner surface, and all light will be totally contained within the pipe until it exits at the far end. At the entrance to the optic fiber, however, the light must strike the end at a high incidence angle in order to travel across the boundary and into the fiber.

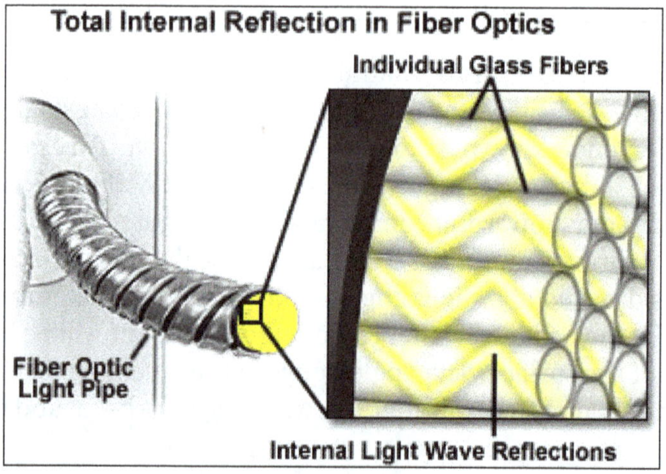

The principles of reflection are exploited to great benefit in many optical instruments and devices, and this often includes the application of various mechanisms to reduce reflections from surfaces that take part in image formation. The concept behind antireflection technology is to control the light used in an optical device in such a manner that the light rays reflect from surfaces where it is intended and beneficial, and do not reflect away from surfaces where this would have a deleterious effect on the image being observed. One of the most significant advances made in modern lens design, whether for microscopes, cameras, or other optical devices, is the improvement in antireflection coating technology.

Thin coatings of certain materials, when applied to lens surfaces, can help reduce unwanted reflections from the surfaces that can occur when light passes through a lens system. Modern lenses that are highly corrected for optical aberrations generally have multiple individual lenses, or lens elements, which are mechanically held together in a barrel or lens tube, and are more properly referred to as a lens or optical system. Each air-glass interface in such a system, if not coated to reduce reflections, can reflect between four and five percent of an incident light beam normal to the surface, resulting in a transmission value of 95 to 96 percent at normal incidence. Application of a quarter-wavelength thick antireflection coating having a specifically chosen refractive index can increase the transmission value by three to four percent.

Modern objective lenses for microscopes, as well as those designed for cameras and other optical devices, have become increasingly more sophisticated and complex, and may have 15 or more separate lens elements with multiple air-glass interfaces. If none of the elements were coated, reflection losses in the lens from axial rays alone would reduce transmittance values to around 50 percent. In the past, single-layer coatings were used to reduce glare and improve light transmission, but these have been largely supplanted by multilayer coatings that can produce transmittance values exceeding 99.9 percent for visible light.

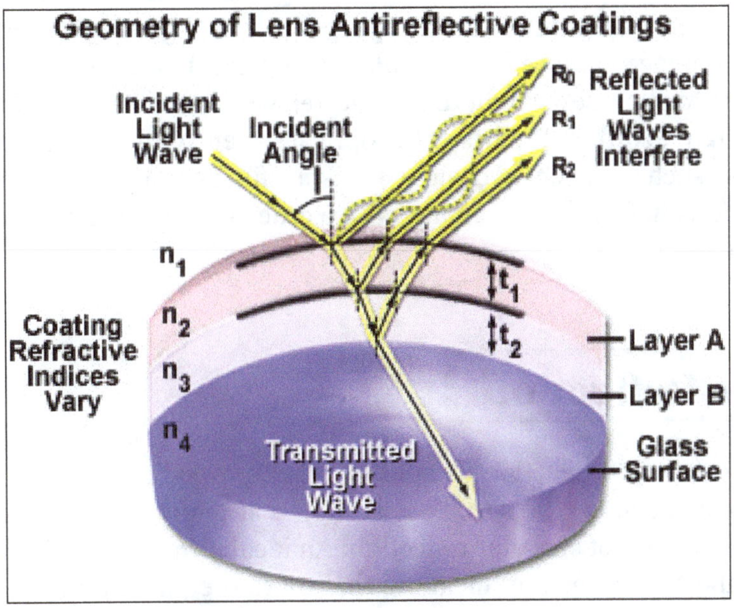

Illustrated in figure is a schematic drawing of light waves reflecting from and/or passing through a lens element coated with two antireflection layers. The incident wave strikes the first layer (Layer A in figure) at an angle, resulting in part of the light being reflected (R(0)) and part being transmitted through the first layer. Upon encountering the second antireflection layer (Layer B), another portion of the light (R(1)) is reflected at the same angle and interferes with light reflected from the first layer. Some of the remaining light waves continue on to the glass surface where they are again partially reflected and partially transmitted. Light that is reflected from the glass surface (R(2)) interferes (both constructively and destructively) with light reflected from the antireflection layers. The refractive indices of the antireflection layers differ from that of the glass and the surrounding medium (air), and are carefully chosen according to the composition of the glass used in the particular lens element to produce the desired refraction angles. As the light waves pass through the antireflection coatings and the glass lens surface, nearly all of the light (depending upon the angle of incidence) is ultimately transmitted through the lens element and focused to form an image.

Magnesium fluoride is one of many materials used for thin-layer optical antireflection coatings, although most microscope and lens manufacturers now produce their own proprietary coating formulations. The general result of these antireflection measures is a dramatic improvement of image quality in optical devices because of increased transmission of visible wavelengths, reduction of glare from unwanted reflections, and elimination of interference from unwanted wavelengths that lie outside the visible light spectral range.

The reflection of visible light is a property of the behavior of light that is fundamental in the function of all modern microscopes. Light is often reflected by one or more plane (or flat) mirrors within the microscope to direct the light path through lenses that form the virtual images we see in the oculars (eyepieces). Microscopes also make use of beamsplitters to allow some light to be reflected while simultaneously transmitting a portion of the light to different parts of the optical system. Other optical components in the microscope, such as specially designed prisms, filters, and lens coatings, also carry out their functions in forming the image with a crucial reliance on the phenomenon of light reflection.

Diffuse Reflection

Diffuse reflection is the reflection of light or other waves or particles from a surface such that a ray incident on the surface is scattered at many angles rather than at just one angle as in the case of specular reflection. An ideal diffuse reflecting surface is said to exhibit Lambertian reflection, meaning that there is equal luminance when viewed from all directions lying in the half-space adjacent to the surface.

A surface built from a non-absorbing powder such as plaster, or from fibers such as paper, or from a polycrystalline material such as white marble, reflects light diffusely with great efficiency. Many common materials exhibit a mixture of specular and diffuse reflection.

The visibility of objects, excluding light-emitting ones, is primarily caused by diffuse reflection of light: it is diffusely-scattered light that forms the image of the object in the observer's eye.

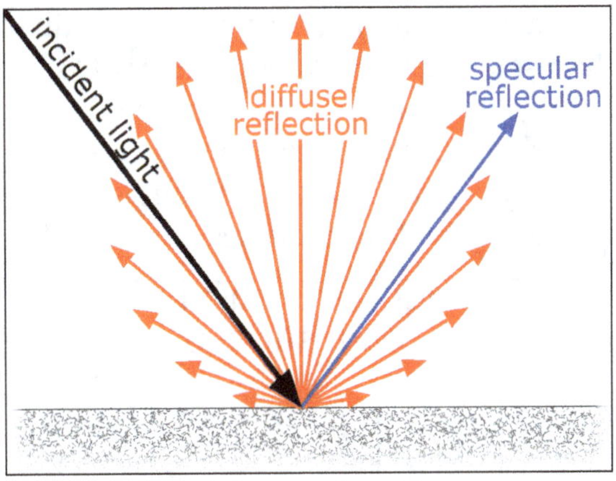

Diffuse and specular reflection from a glossy surface. The rays represent luminous intensity, which varies according to Lambert's cosine law for an ideal diffuse reflector.

Mechanism

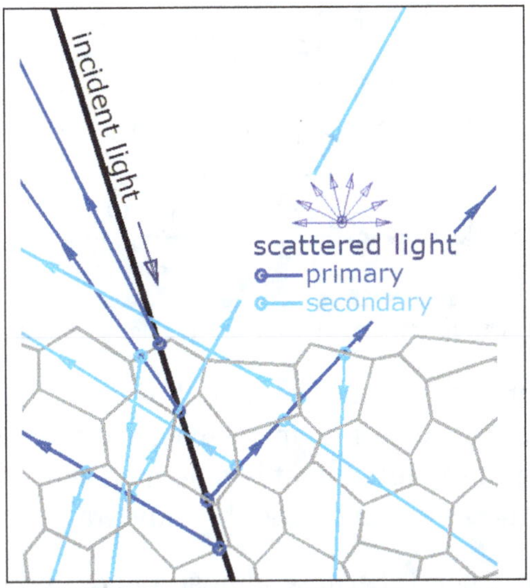

General mechanism of diffuse reflection by a solid surface.

Diffuse reflection from solids is generally not due to surface roughness. A flat surface is indeed required to give specular reflection, but it does not prevent diffuse reflection. A piece of highly polished white marble remains white; no amount of polishing will turn it into a mirror. Polishing produces some specular reflection, but the remaining light continues to be diffusely reflected.

The most general mechanism by which a surface gives diffuse reflection does not involve exactly the surface: most of the light is contributed by scattering centers beneath the surface, as illustrated in figure. If one were to imagine that the figure represents snow, and that the polygons are its (transparent) ice crystallites, an impinging ray is partially reflected (a few percent) by the first particle, enters in it, is again reflected by the interface with the second particle, enters in it, impinges on the third, and so on, generating a series of "primary" scattered rays in random directions, which, in turn, through the same mechanism, generate a large number of "secondary" scattered rays, which generate "tertiary" rays, and so forth. All these rays walk through the snow crystallites, which do not absorb light, until they arrive at the surface and exit in random directions. The result is that the light that was sent out is returned in all directions, so that snow is white despite being made of transparent material (ice crystals).

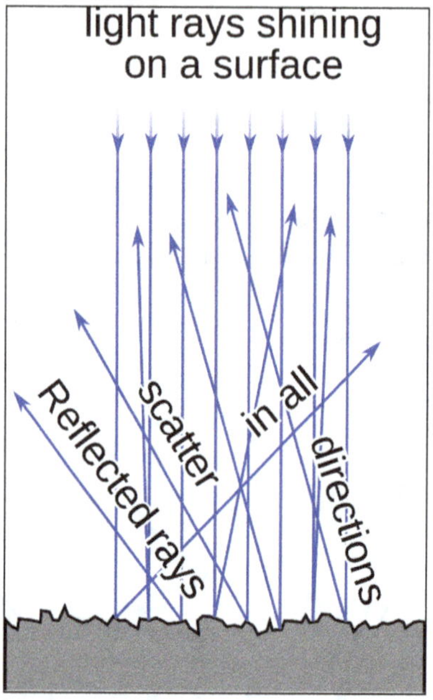

Diffuse reflection from an irregular surface.

For simplicity, "reflections" are spoken of here, but more generally the interface between the small particles that constitute many materials is irregular on a scale comparable with light wavelength, so diffuse light is generated at each interface, rather than a single reflected ray, but the story can be told the same way.

This mechanism is very general, because almost all common materials are made of "small things" held together. Mineral materials are generally polycrystalline: one can describe them as made of a 3D mosaic of small, irregularly shaped defective crystals. Organic materials are usually composed of fibers or cells, with their membranes and their complex internal structure. And each interface, inhomogeneity or imperfection can deviate, reflect or scatter light, reproducing the above mechanism.

Few materials do not cause diffuse reflection: among these are metals, which do not allow light to enter; gases, liquids, glass, and transparent plastics (which have a liquid-like amorphous microscopic structure); single crystals, such as some gems or a salt crystal; and some very special materials, such as the tissues which make the cornea and the lens of an eye. These materials can reflect diffusely, however, if their surface is microscopically rough, like in a frost glass, or, of course, if their homogeneous structure deteriorates, as in cataracts of the eye lens.

A surface may also exhibit both specular and diffuse reflection, as is the case, for example, of glossy paints as used in home painting, which give also a fraction of specular reflection, while matte paints give almost exclusively diffuse reflection.

Most materials can give some specular reflection, provided that their surface can be polished to eliminate irregularities comparable with the light wavelength (a fraction of a micrometer). Depending on the material and surface roughness, reflection may be mostly specular, mostly diffuse, or anywhere in between. A few materials, like liquids and glasses, lack the internal subdivisions which produce the subsurface scattering mechanism described above, and so give only specular reflection. Among common materials, only polished metals can reflect light specularly with high efficiency, as in aluminum or silver usually used in mirrors. All other common materials, even when perfectly polished, usually give not more than a few percent specular reflection, except in particular cases, such as grazing angle reflection by a lake, or the total reflection of a glass prism, or when structured in certain complex configurations such as the silvery skin of many fish species or the reflective surface of a dielectric mirror. Diffuse reflection can be highly efficient, as in white materials, due to the summing up of the many subsurface reflections.

Colored Objects

Up to this point white objects have been discussed, which do not absorb light. But the above scheme continues to be valid in the case that the material is absorbent. In this case, diffused rays will lose some wavelengths during their walk in the material, and will emerge colored.

Diffusion affects the color of objects in a substantial manner because it determines the average path of light in the material, and hence to which extent the various wavelengths are absorbed. Red ink looks black when it stays in its bottle. Its vivid color is only perceived when it is placed on a scattering material (e.g. paper). This is so because light's path through the paper fibers (and through the ink) is only a fraction of millimeter long. However, light from the bottle has crossed several centimeters of ink and has been heavily absorbed, even in its red wavelengths.

And, when a colored object has both diffuse and specular reflection, usually only the diffuse component is colored. A cherry reflects diffusely red light, absorbs all other colors and has a specular reflection which is essentially white (if the incident light is white light). This is quite general, because, except for metals, the reflectivity of most materials depends on their refractive index, which varies little with the wavelength (though it is this variation that causes the chromatic dispersion in a prism), so that all colors are reflected nearly with the same intensity. Reflections from different origin, instead, may be colored: metallic reflections, such as in gold or copper, or interferential reflections: iridescences, peacock feathers, butterfly wings, beetle elytra, or the antireflection coating of a lens.

Importance for Vision

Looking at one's surrounding environment, the vast majority of visible objects are seen primarily by diffuse reflection from their surface. This holds with few exceptions, such as glass, reflective liquids, polished or smooth metals, glossy objects, and objects that themselves emit light: the Sun, lamps, and computer screens (which, however, emit diffuse light). Outdoors it is the same, with perhaps the exception of a transparent water stream or of the iridescent colors of a beetle. Additionally, Rayleigh scattering is responsible for the blue color of the sky, and Mie scattering for the white color of the water droplets of clouds.

Light scattered from the surfaces of objects is by far the primary light which humans visually observe.

Interreflection

Diffuse interreflection is a process whereby light reflected from an object strikes other objects in the surrounding area, illuminating them. Diffuse interreflection specifically describes light reflected from objects which are not shiny or specular. In real life terms what this means is that light is reflected off non-shiny surfaces such as the ground, walls, or fabric, to reach areas not directly in view of a light source. If the diffuse surface is colored, the reflected light is also colored, resulting in similar coloration of surrounding objects.

In 3D computer graphics, diffuse interreflection is an important component of global illumination. There are a number of ways to model diffuse interreflection when rendering a scene. Radiosity and photon mapping are two commonly used methods.

The Laws of Reflection

Light is known to behave in a very predictable manner. If a ray of light could be observed approaching and reflecting off of a flat mirror, then the behavior of the light as it reflects would follow a predictable law known as the law of reflection. The diagram below illustrates the law of reflection.

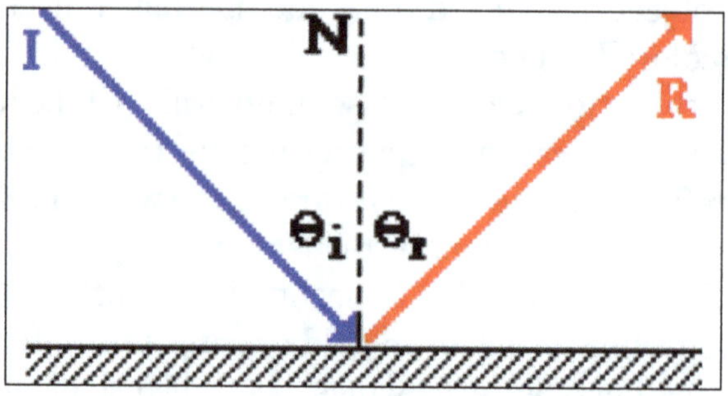

In the diagram, the ray of light approaching the mirror is known as the incident ray (labeled I in the diagram). The ray of light that leaves the mirror is known as the reflected ray (labeled R in the diagram). At the point of incidence where the ray strikes the mirror, a line can be drawn perpendicular to the surface of the mirror. This line is known as a normal line (labeled N in the diagram). The normal line divides the angle between the incident ray and the reflected ray into two equal angles. The angle between the incident ray and the normal is known as the angle of incidence. The angle between the reflected ray and the normal is known as the angle of reflection. (These two angles are labeled with the Greek letter "theta" accompanied by a subscript; read as "theta-i" for angle of incidence and "theta-r" for angle of reflection.) The law of reflection states that when a ray of light reflects off a surface, the angle of incidence is equal to the angle of reflection.

Reflection and the Locating of Images

It is common to observe this law at work in a Physics lab. To view an image of a pencil in a mirror, you must sight along a line at the image location. As you sight at the image, light travels to your eye along the path shown in the diagram below. The diagram shows that the light reflects off the mirror in such a manner that the angle of incidence is equal to the angle of reflection.

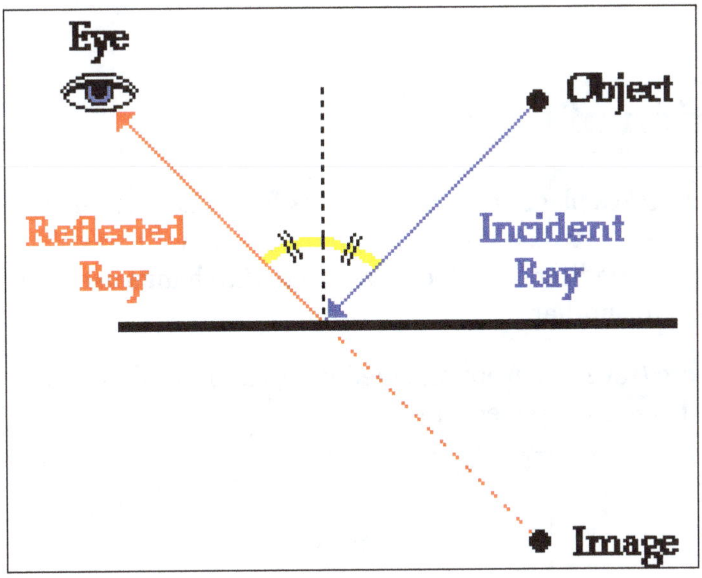

It just so happens that the light that travels along the line of sight to your eye follows the law of reflection. If you were to sight along a line at a different location than the image location, it would be impossible for a ray of light to come from the object, reflect off the mirror according to the law of reflection, and subsequently travel to your eye. Only when you sight at the image, does light from the object reflect off the mirror in accordance with the law of reflection and travel to your eye.

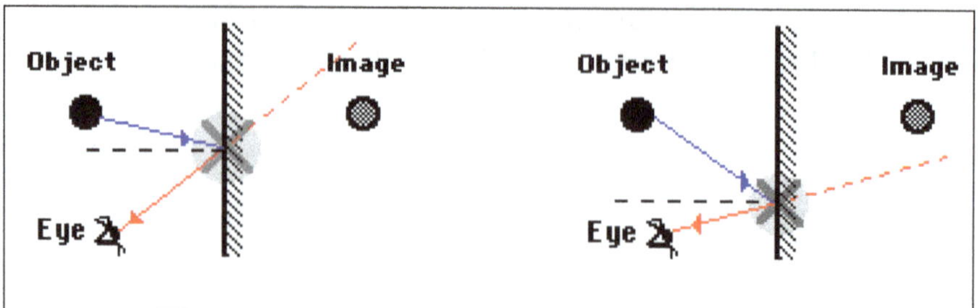

Diagram A Diagram B

For example, in Diagram A above, the eye is sighting along a line at a position *above* the actual image location. For light from the object to reflect off the mirror and travel to the eye, the light would have to reflect in such a way that the angle of incidence is less than the angle of reflection. In Diagram B above, the eye is sighting along a line at a position *below* the actual image location. In this case, for light from the object to reflect off the mirror and travel to the eye, the light would have to reflect in such a way that the angle of incidence is more than the angle of reflection. Neither of these cases would follow the law of reflection. In fact, in each case, the image is not seen when sighting along the indicated line of sight. It is because of the law of reflection that an eye must sight at the image location in order to see the image of an object in a mirror.

Refraction of Light

Refraction of light takes place when light travels from one medium to another. It takes place at the boundary between the two mediums. Also, we know that speed of light is different in different mediums. So, it occurs due to the change in speed of light on going from one medium to another.

If the light rays are travelling from one medium to another they change their direction at the boundary between two mediums.

When the light rays either bend or change their direction while passing from one medium to another it is called refraction of light. The refraction of light takes place when light travels from air into glass, from glass into air, from air into water or from water into air.

The example of optical instruments that work on the basis of refraction of light are camera, microscope etc.

- Incident ray: The light rays passing from air into glass or water are called incident rays.
- Refracted ray: When the light rays bend after passing into another medium, they are called refracted rays.
- Normal: The point of incidence is called normal.
- Angle of incidence: The angel between incident ray and normal is called angel of incidence.
- Angle of refraction: The angel between refracted ray and normal is called angle of refraction.

The angle of refraction is either smaller or greater than angle of incidence.

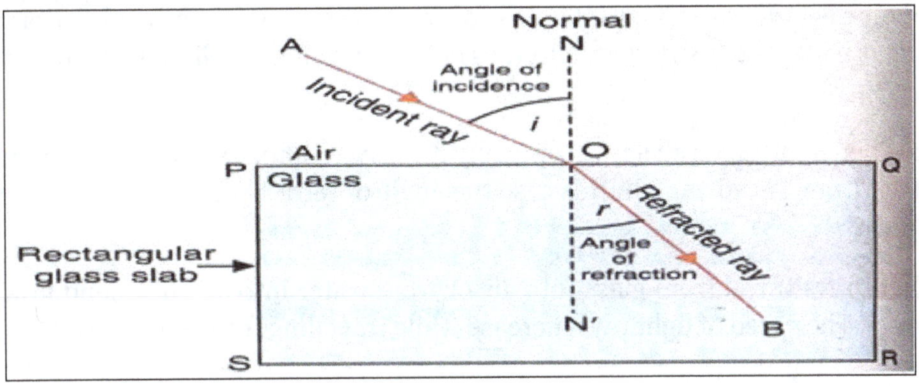

Causes of Refraction

Light travels in different speed in different mediums. For example light travels faster in air than in a glass. Therefore, it is due to the change of speed of light in different medium that the light rays are refracted.

Optically Rarer Medium

A transparent substance (medium) in which the speed of light is more is called optically rarer medium.

Optically Denser Medium

A transparent substance (medium) in which the speed of light is less is known as optically denser medium.

Glass is an optically denser medium than air and water.

Rules for Refraction of Light

Case 1: When light rays travel from optically rarer medium to denser medium then they bend towards normal. In this case angle of refraction is smaller than angel of incidence.

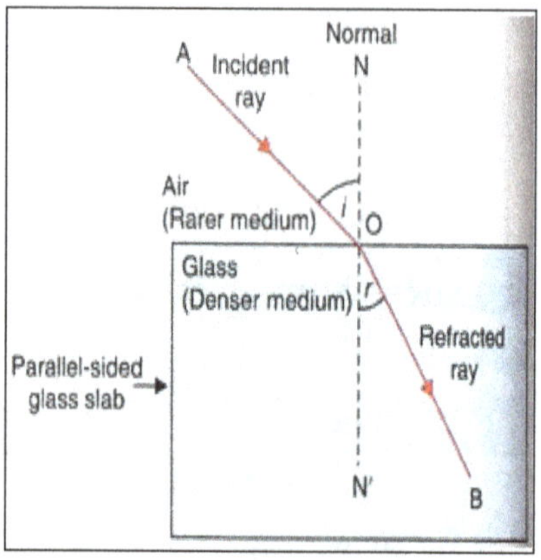

When light rays travel from air into glass or from air into water, it bends towards normal. This is because the speed of light rays decrease while travelling from air into glass or water.

Case 2: When light rays travel from optically denser medium to rarer medium then they bend away from the normal. In this case the angle of refraction is greater than angle of incidence.

When light rays travel from glass into air or from water into air they bend away from the normal. The speed of light rays increase while travelling from glass or water into air.

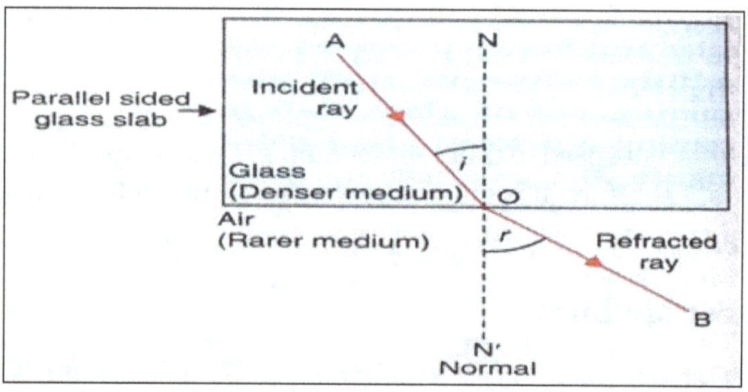

Case of Light going from Air into Glass and again into Air

In this case refraction of light takes place two times. One when it enters the glass slab from air and second time when it enters the air through glass slab.

When light rays travelling through air enters glass slab, they get refracted and bend towards the normal. Now the direction of refracted ray changes again when it comes out of the glass slab into air. Since the ray of light I know travelling from denser medium to rarer medium, it bends away from the normal.

In this case incident ray and the emergent ray are parallel to each other. The perpendicular distance between the original path of incident ray and the emergent ray coming out of the glass slab is called lateral displacement of the emergent ray of light and the angle which the emergent ray makes with the normal is called the angle of emergence.

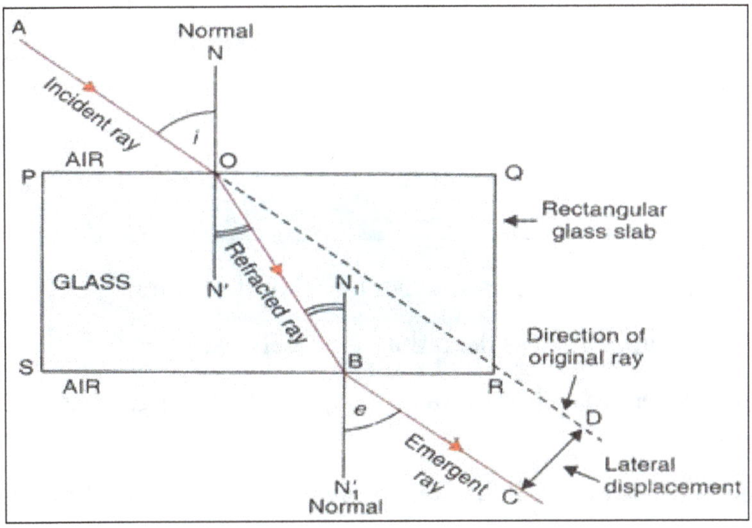

Light Falling Perpendicularly on Glass Slab

When light falls perpendicularly or normally on the surface of a glass slab, it goes straight. There is no bending of ray of light on entering the glass slab or coming out of it. In this case angle of incidence and angle of refraction is zero.

The same happens if the ray of light falls perpendicularly on the surface of water.

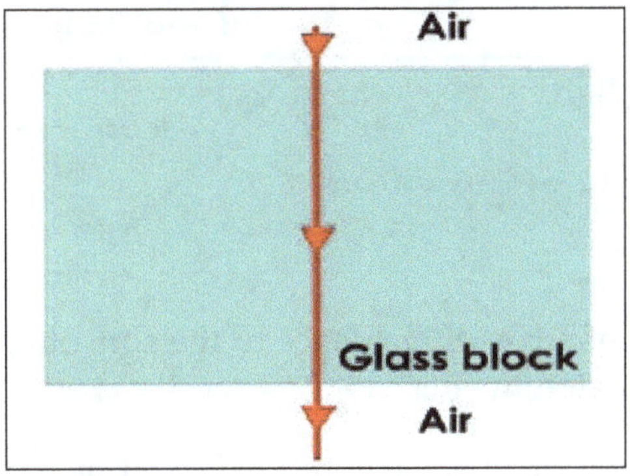

Effects of Refraction of Light

- It is due to refraction of light that when we hold a stick obliquely and partially immersed in water it appears to be bend at the surface of water.

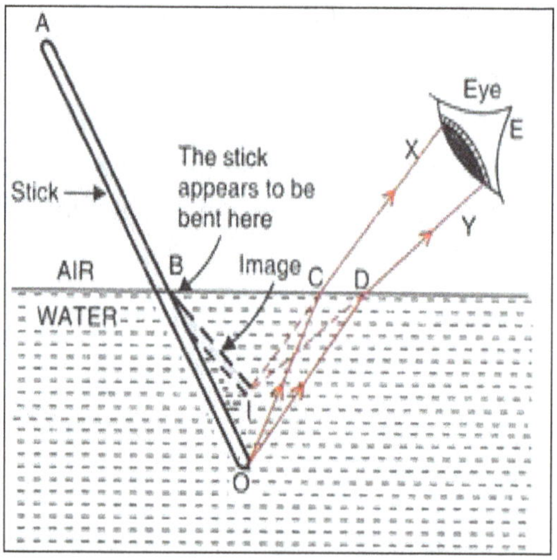

- An object appears to be raised when paced under water.
- Pool of water appears less deep than it actually is.
- If a lemon is kept in a glass of water it appears to be bigger when viewed from the sides of glass.
- It is due to refraction of light that stars appear to twinkle at night.

Laws of Refraction of Light

- The incident ray, refracted ray and normal at the point of incidence, all lie in the same plane, i.e. the surface.

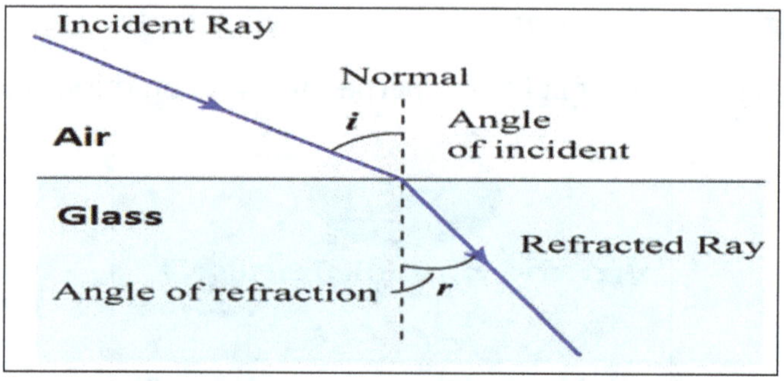

- The ratio of sine of angle of incidence to the sine of angle of refraction is constant for a given pair of media.

Sine of angle of incidence/sine of angle of refraction = Constant

Constant is called refractive index.

Or Sine i/Sine r = constant

The refractive index of a medium helps to know the light-bending ability of that medium.

Refractive Index and Speed of Light

Refractive index of medium 2 with respect to medium 1 is equal to the ratio of speed of light in medium 1 to the speed of light in medium 2.

Relative Refractive Index

When light travels from one medium to another other than vacuum and air, then the value of refractive index is called relative refractive index.

Refractive Index = Speed of light in vacuum/ Speed of light in medium

Or Refractive index = Speed of light in medium 1/ Speed of Light in medium 2

For example, light travelling from water into glass.

Absolute Refractive Index

When light travels from vacuum to another medium, it is called absolute refractive index.

The substance that has higher refractive index is optically denser than another substance having lower refractive index.

Also, the refractive index for light going from medium 1 to medium 2 is equal to the reciprocal of the refractive index of light going from medium to 2 to medium 1.

The Laws of Refraction

The law of refraction, which is generally known as Snell's law, governs the behaviour of light-rays as they propagate across a sharp interface between two transparent dielectric media.

Consider a light-ray incident on a plane interface between two transparent dielectric media, labelled 1 and 2, as shown in figure. The law of refraction states that the incident ray, the refracted ray, and the normal to the interface, all lie in the same plane. Furthermore,

$$n_1 \sin \theta_1 = n_2 \sin \theta_2$$

where θ_1 is the angle subtended between the incident ray and the normal to the interface, and θ_2 is the angle subtended between the refracted ray and the normal to the interface. The quantities n_1 and n_2 are termed the refractive indices of media 1 and 2, respectively. Thus, the law of refraction predicts that a light-ray always deviates more towards the normal in the optically denser medium: i.e., the medium with the higher refractive index. Note that $n_2 > n_1$ in the figure. The law of refraction also holds for non-planar interfaces, provided that the normal to the interface at any given point is understood to be the normal to the local tangent plane of the interface at that point.

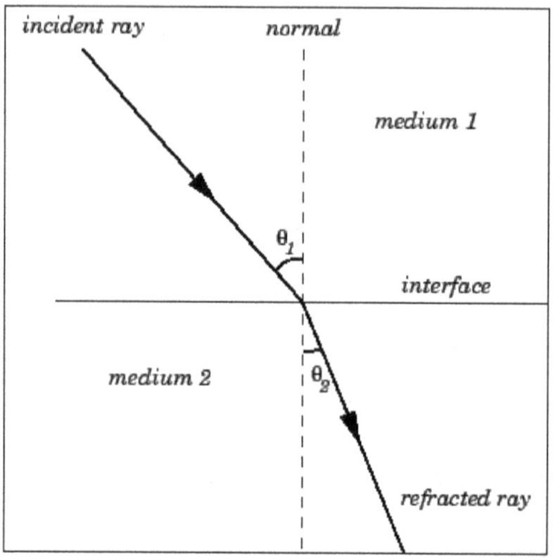

The law of refraction.

By definition, the refractive index n of a dielectric medium of dielectric constant K is given by

$$n = \sqrt{K}.$$

Table shows the refractive indices of some common materials (for yellow light of wave-length $\lambda = 589$ nm).

Table: Refractive indices of some common materials at $\lambda = 589$ nm.

Material	n
Air (STP)	1.00029
Water	1.33
Ice	1.31
Glass:	
Light flint	1.58
Heavy flint	1.65
Heaviest flint	1.89
Diamond	2.42

The law of refraction follows directly from the fact that the speed u with which light propagates through a dielectric medium is inversely proportional to the refractive index of the medium. In fact,

$$v = \frac{c}{n},$$

where c is the speed of light in a vacuum. Consider two parallel light-rays, *a* and *b*, incident at an angle θ_1 with respect to the normal to the interface between two dielectric media, 1 and 2. Let the refractive indices of the two media be n_1 and n_2 respectively, with $n_2 > n_1$. It is clear from figure. that ray *b* must move from point *B* to point *Q*, in medium 1, in the same time interval, Δt, in which ray *a* moves between points *A* and *P*, in medium 2. Now, the speed of light in medium 1 is $v_1 = c / n_1$, whereas the speed of light in medium 2 is $v_2 = c / n_2$. It follows that the length *BQ* is given by $v_1 \Delta t$, whereas the length *AP* is given by $v_2 \Delta t$. By trigonometry,

$$\sin \theta_1 = \frac{BQ}{AQ} = \frac{v_1 \Delta t}{AQ},$$

and,

$$\sin \theta_2 = \frac{AP}{AQ} = \frac{v_2 \Delta t}{AQ}.$$

Hence,

$$\frac{\sin \theta_1}{\sin \theta_2} = \frac{v_1}{v_2} = \frac{n_2}{n_1},$$

which can be rearranged to give Snell's law. Note that the lines AB and PQ represent wave-fronts in media 1 and 2, respectively, and, therefore, cross rays a and b at right-angles.

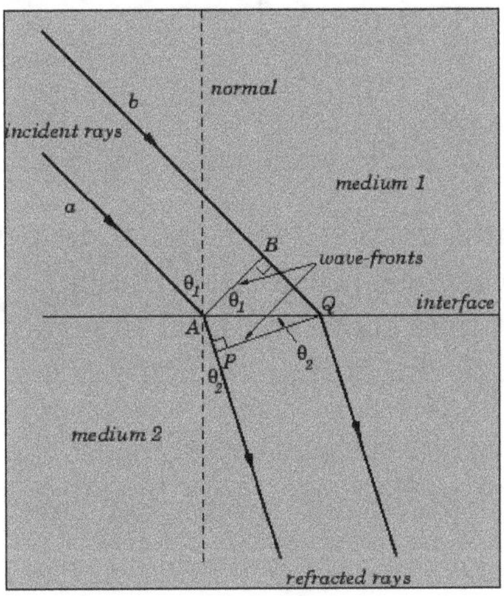

Derivation of Snell's law.

When light passes from one dielectric medium to another its velocity u changes, but its frequency f remains unchanged. Since, $v = f\lambda$ for all waves, where λ is the wave-length, it follows that the wavelength of light must also change as it crosses an interface between two different media. Suppose that light propagates from medium 1 to medium 2. Let n_1 and n_2 be the refractive indices of the two media, respectively. The ratio of the wave-lengths in the two media is given by:

$$\frac{\lambda_2}{\lambda_1} = \frac{v_2/f}{v_1/f} = \frac{v_2}{v_1} = \frac{n_1}{n_2}.$$

Thus, as light moves from air to glass its wavelength decreases.

Refraction at Spherical Surfaces

Focus

Consider a convex circular boundary (of radius R and center of curvature O) between two regions.

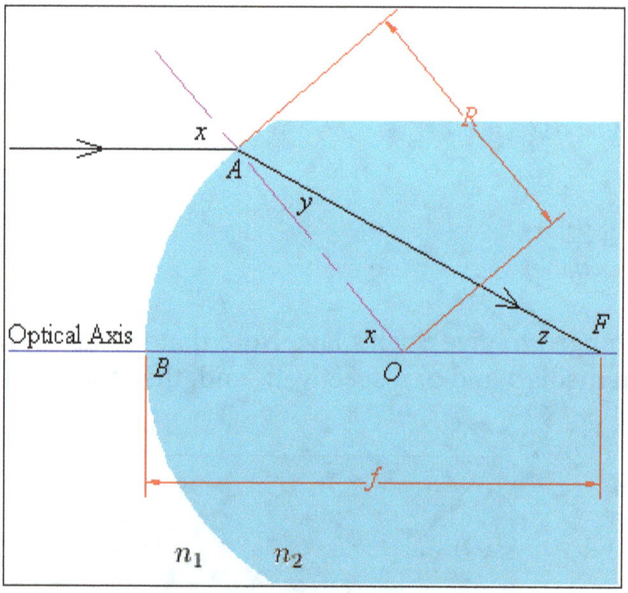

If rays coming in from infinity are close to the optical axis, then we will be dealing with small angles, and the small angle approximation $\sin(w) = w$ will be good. We assume that the angles x, y, z as labelled above are small enough so the approximation holds, and we now proceed to find the focus of this setup.

When the ray hits the point A, it refracts according to Snell's Law; using small angle approximations, this reduces to $n_1 x = n_2 y$. Also, the angle sum theorems for lines and triangles yield $z = x - y$.

Using radian measure, AB = Rx, and the relationship $AB = fz = f(x-y)$ holds approximately. Hence, keeping in mind our simplified Snell's Law, we have:

$$f \approx \frac{AB}{x-y} = \frac{Rx}{x-y}$$

$$= \frac{R\frac{n_2 y}{n_1}}{\frac{n_2 y}{n_1} - y} = \frac{Rn_2 n_1}{n_1(n_2 - n_1)}$$

$$= \frac{Rn_2}{n_2 - n_1}$$

The same equation holds for a concave boundary; for brevity's sake we shall omit the proof. In this case, the focus is the spot from which the refracted rays seem to diverge.

Image of a Point

In discussing refraction, we can use the same ray tracing techniques that we used for mirrors. Here, we consider a ray that is bent to intersect the focus, and a ray through the centre of curvature (which is therefore not bent).

Observe the following convex boundary.

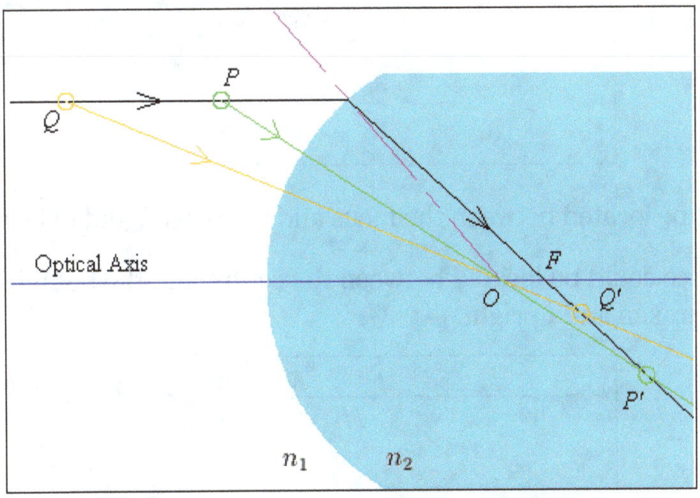

We see that the image is inverted and shrunken, and that it is located in the second region.

As with concave mirrors, concave refractive boundaries have three cases.

When the source is beyond the focus, we get a diminished, upright image in front of the source, located between the focus and the centre of curvature.

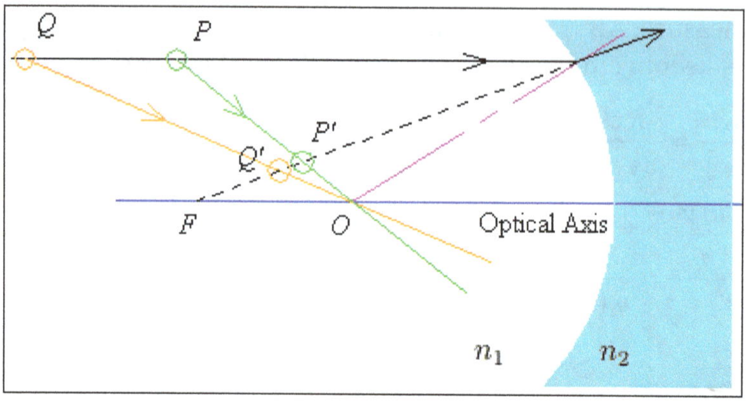

The image must be between the focus and the centre because we are travelling from an area of low index to one with a high index; the refracted ray cannot be bent past the normal line.

When the source is between the focus and the centre of curvature, the image is again diminished and upright.

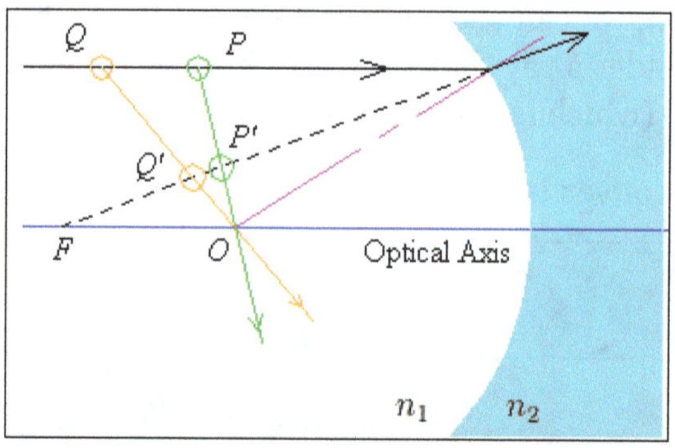

The image must be located between the focus and the centre, and in front of the source.

Finally, the source could be located between the centre and the boundary. In this case, the image is shrunken and upright.

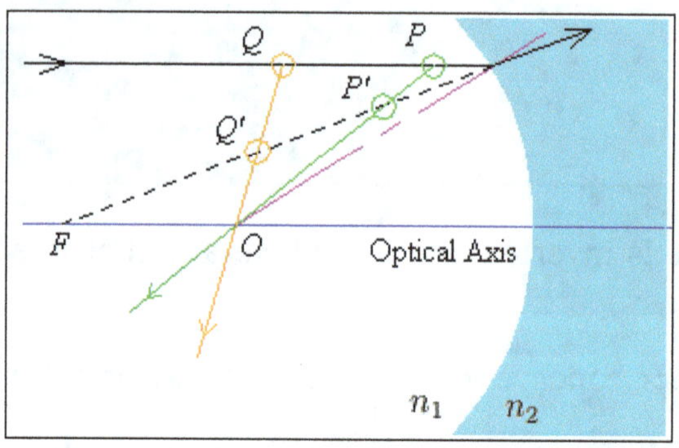

The image will be located between the boundary and the centre; it will also be behind the source. This is an easy consequence of the geometry; since the source is in front of the centre, the ray that goes through the centre must go backwards to intersect the line that goes through the focus.

Precise Location of the Image

The image below provides a method for finding the image of an object after refraction at a convex boundary.

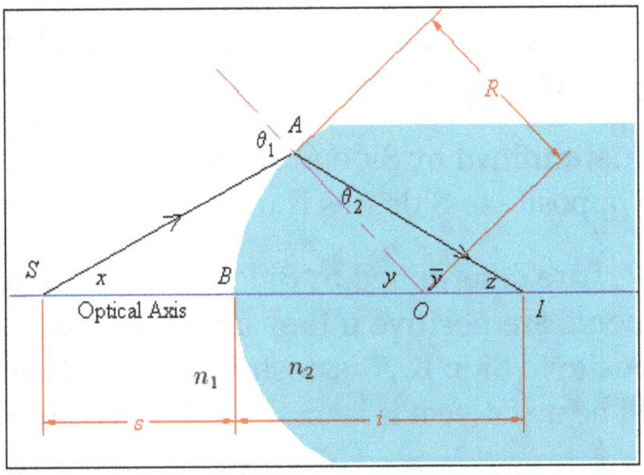

The source is located at S, and the image will be located at I. The centre of curvature is at O, and the radius is R. The angle x can presumably be measured, and the angles of incidence and refraction are related using our small angle approximation of Snell's Law. Using the angle sum theorems for lines and triangles, and radian measure, we find that:

$$\theta_1 = x + y$$
$$\theta_2 = \pi - \bar{y} - z = \pi - (\pi - y) - z = y - z$$

so that our small angle Snell's Law is given by:

$$n_1(x+y) = n_2(y-z)$$

Using radians, the arclength $AB = Ry$. For small angles, $AB = sx = iz$ holds. Substituting these formulas into our Snellian approximation gives:

$$n_1\left(\frac{AB}{s} + \frac{AB}{R}\right) = n_2\left(\frac{AB}{R} - \frac{AB}{i}\right)$$
$$\Rightarrow \frac{n_1}{s} + \frac{n_2}{i} = \frac{n_2 - n_1}{R}$$

In fact, the equation holds for concave boundaries as well, provided we use sign conventions for our measurements s, i, R.

Sign Conventions

In our presentation of mirrors, we spoke of sign conventions for measurements that allowed us to use the same formulas for all spherical mirrors. A modified version of these conventions actually allows us to deal with both reflecting and refracting surfaces in a general manner.

Let Side A of an optical component be the side from which light starts, and let Side B be the side to which light travels. With mirrors, Sides A and B are identical. If s is the source distance, i is the image distance, R is the radius of curvature, and f is the focal length, then, our sign conventions are as follows.

- The sign of s is determined by Side A: If the source is on Side A, s is positive; if it is on the side opposite to Side A, s is negative.
- The signs of i, R, f are determined by Side B: For the image and the focal point, their measurements are positive if they are on Side B, and negative if they are on the side opposite to Side B. R is positive provided the centre of curvature is on Side B; it will be negative if it is on the side opposite Side B.

It cannot be stressed enough that for a mirror, Sides A and B are the same.

Total Internal Reflection

Total Internal Reflection (TIR) is the phenomenon that makes the water-to-air surface in a fish-tank look like a perfectly silvered mirror when viewed from below the water level. Technically, TIR is the total reflection of a wave incident at a sufficiently oblique angle on the interface between two media, of which the second ("external") medium is transparent to such waves but has a higher wave velocity than the first ("internal") medium. TIR occurs not only with electromagnetic waves such as light waves and microwaves, but also with other types of waves, including sound and water waves. In the case of a narrow train of waves, such as a laser beam, we tend to speak of the total internal reflection of a "ray".

The color of the laser light itself is deep violet; but its wavelength is short enough to cause fluorescence in the glass, which re-radiates greenish light in all directions, rendering the zigzag beam visible.

Refraction is generally accompanied by partial reflection. When a wavetrain is refracted from a medium of lower propagation speed (higher refractive index) to a medium of higher propagation speed (lower refractive index), the angle of refraction (between the refracted ray and the normal to the refracting interface) is greater than the angle of incidence (between the incident ray and the normal to the interface). Hence, as the angle of incidence approaches a certain limit, called the critical angle, the angle of refraction approaches 90°, at which the refracted ray becomes tangential to the interface. As the angle of incidence increases beyond the critical angle, the conditions of refraction can no longer be satisfied; so we have no refracted ray, and the partial reflection becomes total. In an isotropic medium such as air, water, or glass, the ray direction is simply the direction normal to the wavefront.

Underwater plants in an aquarium, and their inverted images (top) formed by total internal reflection in the water-air surface.

Repeated total internal reflection of a 405 nm laser beam between the front and back surfaces of a glass pane.

If the internal and external media are isotropic with refractive indices n_1 and n_2 respectively, the critical angle is given by $\theta_c = \arcsin(n_2 / n_1)$, and is defined if $n_2 \leq n_1$. For example, for visible light, the critical angle is about 49° for incidence from water to air, and about 42° for incidence from common glass to air.

Details of the mechanism of TIR give rise to more subtle phenomena. While total reflection, by definition, involves absolutely no continuing transfer of power across the interface, the external medium carries a so-called evanescent wave, which travels along the interface with an amplitude that falls off exponentially with distance from the interface. The "total" reflection is indeed total if the external medium is lossless (perfectly transparent), continuous, and of infinite extent, but can be conspicuously less than total if the evanescent wave is absorbed by a lossy external medium ("attenuated total reflectance"), or diverted by the outer boundary of the external medium or by objects embedded in that medium ("frustrated" TIR). Unlike partial reflection between transparent media, total internal reflection is accompanied by a non-trivial phase shift (not just zero or 180°) for each component of polarization (normal or parallel to the plane of incidence), and the shifts vary with the angle of incidence. The explanation of this effect by Augustin-Jean Fresnel, in 1823, added to the evidence in favor of the wave theory of light.

The phase shifts are utilized by Fresnel's invention, the Fresnel rhomb, to modify polarization. The efficiency of the reflection is exploited by optical fibers (used in telecommunications cables and in image-forming fiberscopes), and by reflective prisms, such as erecting prisms for binoculars.

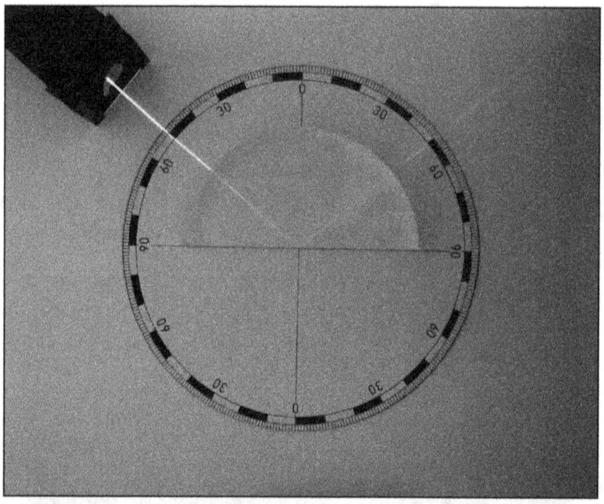

Total internal reflection of light in a semicircular acrylic block.

Although total internal reflection can occur with any kind of wave that can be said to have oblique incidence, including (e.g.) microwaves and sound waves, it is most familiar in the case of light waves.

Total internal reflection of light can be demonstrated using a semicircular-cylindrical block of common glass or acrylic glass. In figure, a "ray box" projects a narrow beam of light (a "ray") radially inward. The semicircular cross-section of the glass allows the incoming ray to remain perpendicular to the curved portion of the air/glass surface, and thence to continue in a straight line towards the flat part of the surface, although its angle with the flat part varies.

Where the ray meets the flat glass-to-air interface, the angle between the ray and the normal to the interface is called the angle of incidence. If this angle is sufficiently small, the ray is partly reflected but mostly transmitted, and the transmitted portion is refracted away from the normal, so that the angle of refraction (between the refracted ray and the normal to the interface) is greater than the angle of incidence. For the moment, let us call the angle of incidence θ_i and the angle of refraction θ_t (where t is for transmitted, reserving r for reflected). As θ_i increases and approaches a certain "critical angle", denoted by θ_c (or sometimes θ_{cr}), the angle of refraction approaches 90° (that is, the refracted ray approaches a tangent to the interface), and the refracted ray becomes fainter while the reflected ray becomes brighter. As θ_i increases beyond θ_c, the refracted ray disappears and only the reflected ray remains, so that all of the energy of the incident ray is reflected; this is total internal reflection (TIR). In brief:

- If $\theta_i < \theta_c$, the incident ray is split, being *partly* reflected and partly refracted;
- If $\theta_i > \theta_c$, the incident ray suffers total internal reflection (TIR); none of it is transmitted.

Critical Angle

The critical angle is the smallest angle of incidence that yields total reflection. For light waves and other electromagnetic waves in isotropic media, there is a well-known formula for the critical angle in terms of the refractive indices. For some other types of waves, it is more convenient to think in terms of propagation velocities rather than refractive indices. The latter approach is more direct and more general, and will therefore be discussed first.

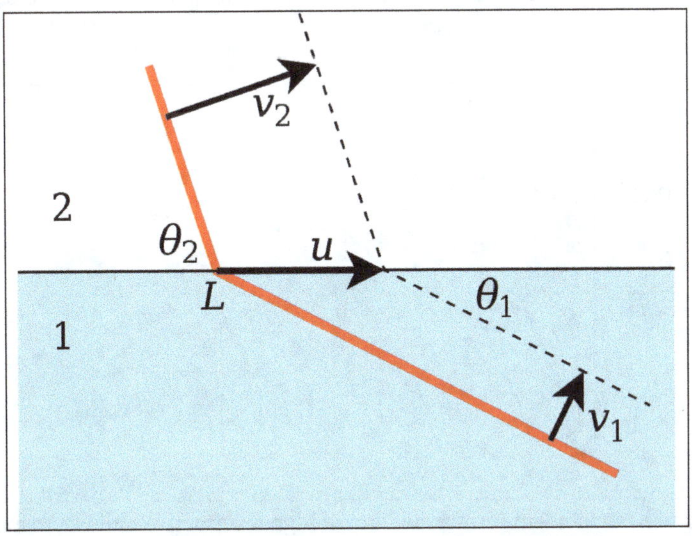

Refraction of a wavefront (red) from a medium with lower normal velocity v_1 to a medium with higher normal velocity v_2. The incident and refracted segments of the wavefront meet in a common line L (seen "end-on"), which travels at velocity u.

When a wavefront is refracted from one medium to another, the incident (incoming) and refracted (outgoing) portions of the wavefront meet at a common line on the refracting surface (interface). Let this line, denoted by *L*, move at velocity *u* across the surface, where u is measured normal to L. Let the incident and refracted wavefronts propagate with normal velocities v_1 and v_2 (respectively), and let them make the dihedral angles θ_1 and θ_2 (respectively) with the interface. From the geometry, v_1 is the component of u in the direction normal to the incident wave, so that $v_1 = u\sin\theta_1$. Similarly, $v_2 \quad u\sin\ _2$. Solving each equation for 1/u and equating the results, we obtain the general law of refraction for waves:

$$\frac{\sin\theta_1}{v_1} = \frac{\sin\theta_2}{v_2}.$$

But the dihedral angle between two planes is also the angle between their normals. So θ_1 is the angle between the normal to the incident wavefront and the normal to the interface, while θ_2 is the angle between the normal to the refracted wavefront and the normal to the interface; and equation $\frac{\sin\theta_1}{v_1} = \frac{\sin\theta_2}{v_2}$ tells us that the sines of these angles are in the same ratio as the respective velocities.

This result has the form of "Snell's law", except that we have not yet said that the ratio of velocities is constant, nor identified θ_1 and θ_2 with the angles of incidence and refraction (called θ_i and θ_t above). However, if we now suppose that the media are *isotropic*, two further conclusions follow: first, the two velocities, and hence their ratio, are independent of their directions; and second, the wave-normal directions coincide with the *ray* directions, so that θ_1 and θ_2 coincide with the angles of incidence and refraction as defined above.

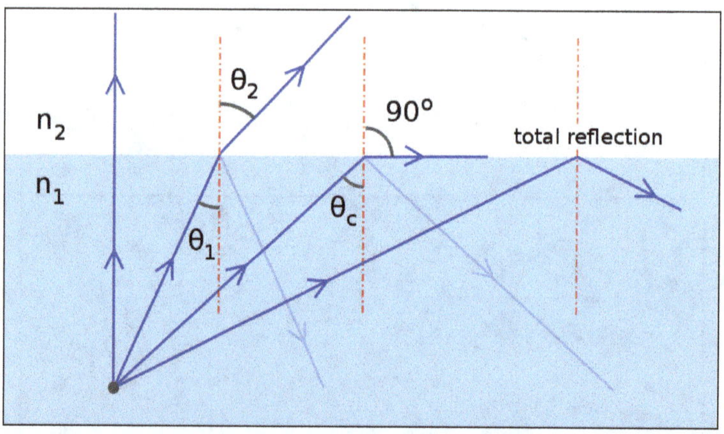

Behavior of a ray incident from a medium of higher refractive index n_1 to a medium of lower refractive index n_2, at increasing angles of incidence.

Obviously the angle of refraction cannot exceed 90°. In the limiting case, we putθ_2 = 90° and$\theta_1 = \theta_c$ in equation ($\frac{\sin\theta_1}{v_1} = \frac{\sin\theta_2}{v_2}$), and solve for the critical angle:

$$\theta_c = \arcsin(v_1 / v_2)$$

In deriving this result, we retain the assumption of isotropic media in order to identify θ_1 and θ_2 with the angles of incidence and refraction.

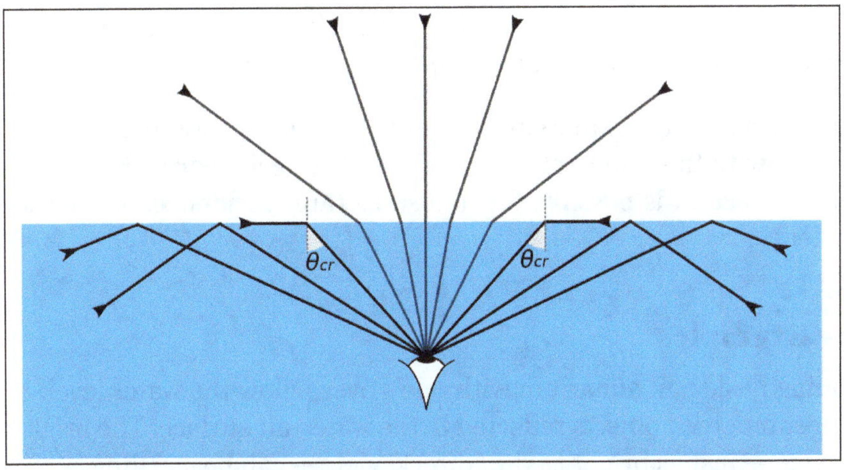

The angle of refraction for grazing incidence from air to water is the critical angle for incidence from water to air.

For electromagnetic waves, and especially for light, it is customary to express the above results in terms of refractive indices. The refractive index of a medium with normal velocity v_1 is defined as $n_1 = c / v_1$, where c is the speed of light in a vacuum. Hence $v_1 = c / n_1$. Similarly, $v_2 = c / n_2$. Making these substitutions in equations ($\frac{\sin\theta_1}{v_1} = \frac{\sin\theta_2}{v_2}$.) and ($\theta_c = \arcsin(v_1 / v_2)$), we obtain:

$$n_1 \sin\theta_1 = n_2 \sin\theta_2$$

and,

$$\theta_c = \arcsin(n_2 / n_1)$$

Equation ($n_1 \sin\theta_1 = n_2 \sin\theta_2$) is the law of refraction for general media, in terms of refractive indices, provided that θ_1 and θ_2 are taken as the dihedral angles; but if the media are *isotropic*, then n_1 and n_2 become independent of direction while θ_1 and θ_2 may be taken as the angles of incidence and refraction for the rays, and equation ($\theta_c = \arcsin(n_2 / n_1)$) follows. So, for isotropic media, Eqs. ($n_1 \sin\theta_1 = n_2 \sin\theta_2$) and ($\theta_c = \arcsin(n_2 / n_1)$) together describe the behavior in figure.

According to Eq. ($\theta_c = \arcsin(n_2 / n_1)$), for incidence from water ($n_1 \approx 1.333$) to air ($n_2 \approx 1$), we have $\theta_c \approx 48.6°$, whereas for incidence from common glass or acrylic ($n_1 \approx 1.50$) to air ($n_2 \approx 1$), we have $\theta_c \approx 41.8°$.

The arcsin function yielding θ_c is defined only if $n_2 \leq n_1$ $(v_2 \geq v_1)$. Hence, for isotropic media, total internal reflection cannot occur if the second medium has a higher refractive index (lower normal velocity) than the first. For example, there cannot be TIR for incidence from air to water; rather, the critical angle for incidence from water to air is the angle of refraction at grazing incidence from air to water.

The medium with the higher refractive index is commonly described as optically *denser*, and the one with the lower refractive index as optically *rarer*. Hence it is said that total internal reflection is possible for "dense-to-rare" incidence, but not for "rare-to-dense" incidence.

Everyday Examples

When standing beside an aquarium with one's eyes below the water level, one is likely to see fish or submerged objects reflected in the water-air surface. The brightness of the reflected image – just as bright as the "direct" view – can be startling.

Total internal reflection by the water's surface at the shallow end of a swimming pool.

The broad bubble-like apparition between the swimmer and her reflection is merely a disturbance of the reflecting surface. Some of the space above the water level can be seen through "Snell's window" at the top of the frame.

A similar effect can be observed by opening one's eyes while swimming just below the water's surface. If the water is calm, the surface outside the critical angle (measured from the vertical) appears mirror-like, reflecting objects below. The region above the water cannot be seen except overhead, where the hemispherical field of view is compressed into a conical field known as Snell's window, whose angular diameter is twice the critical angle. The field of view above the water is theoretically 180° across, but seems less because as we look closer to the horizon, the vertical dimension is more strongly compressed by the refraction; e.g., by Eq. ($n_1 \sin\theta_1 = n_2 \sin\theta_2$), for air-to-water incident angles of 90°, 80°, and 70°, the corresponding angles of refraction are 48.6° (θ_{cr} in figure), 47.6°, and 44.8°, indicating that the image of a point 20° above the horizon is 3.8° from the edge of Snell's window while the image of a point 10° above the horizon is only 1° from the edge.

A round "brilliant"-cut diamond.

figure, for example, is a photograph taken near the bottom of the shallow end of a swimming pool. What looks like a broad horizontal stripe on the right-hand wall consists of the lower edges of a row of orange tiles, and their reflections; this marks the water level, which can then be traced across the other wall. The swimmer has disturbed the surface above her, scrambling the lower half of her reflection, and distorting the reflection of the ladder (to the right). But most of the surface is still calm, giving a clear reflection of the tiled bottom of the pool. The space above the water is not visible except at the top of the frame, where the handles of the ladder are just discernible above the edge of Snell's window.

The critical angle influences the angles at which gemstones are cut. The round "brilliant" cut, for example, is designed to refract light incident on the front facets, reflect it twice by TIR off the back facets, and transmit it out again through the front facets, so that the stone looks bright. Diamond is especially suitable for this treatment, because its high refractive index (about 2.42) and consequently small critical angle (about 24.5°) yield the desired behavior over a wide range of viewing angles. Cheaper materials that are similarly amenable to this treatment include cubic zirconia (index ≈ 2.15) and moissanite (non-isotropic, hence doubly refractive, with an index ranging from about 2.65 to 2.69, depending on direction and polarization); both of these are therefore popular as diamond simulants.

Applications

Optical fibers exploit total internal reflection to carry signals over long distances with little attenuation. They are used in telecommunication cables, and in image-forming fiberscopes such as colonoscopes.

In the catadioptric Fresnel lens, invented by Augustin-Jean Fresnel for use in lighthouses, the outer prisms use TIR to deflect light from the lamp through a greater angle than would be possible with purely refractive prisms, but with less absorption of light (and less risk of tarnishing) than with conventional mirrors.

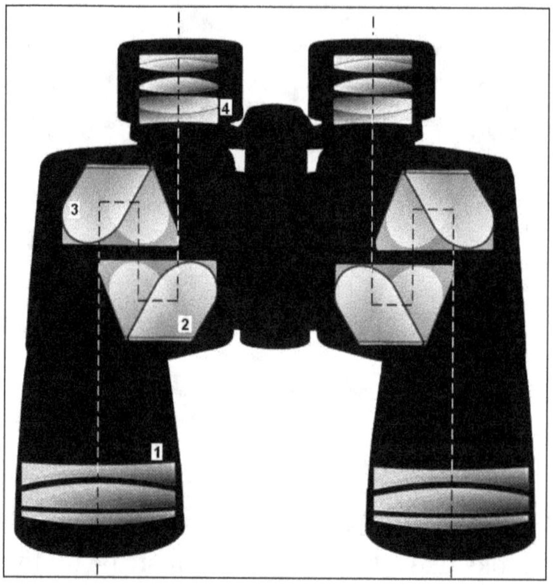

Porro prisms (labeled 2 & 3) in a pair of binoculars.

Other reflecting prisms that use TIR include the following (with some overlap between the categories):

- Image-erecting prisms for binoculars and spotting scopes include paired 45°-90°-45° Porro prisms, the Porro–Abbe prism, the inline Koenig and Abbe–Koenig prisms, and the compact inline Schmidt–Pechan prism. (The last consists of two components, of which one is a kind of Bauernfeind prism, which requires a reflective coating on one of its two reflecting faces, due to a sub-critical angle of incidence.) These prisms have the additional function of folding the optical path from the objective lens to the prime focus, reducing the overall length for a given primary focal length.
- A prismatic star diagonal for an astronomical telescope may consist of a single Porro prism (configured for a single reflection, giving a mirror-reversed image) or an Amici roof prism (which gives a non-reversed image).
- Roof prisms use TIR at two faces meeting at a sharp 90° angle. This category

includes the Koenig, Abbe–Koenig, Schmidt–Pechan, and Amici types (already mentioned), and the roof pentaprism used in SLR cameras; the last of these requires a reflective coating on one non-TIR face.

- A prismatic corner reflector uses three total internal reflections to reverse the direction of incoming light.
- The Dove prism gives an inline view with mirror-reversal.

Polarizing prisms: Although the Fresnel rhomb, which converts between linear and elliptical polarization, is not birefringent (doubly refractive), there are other kinds of prisms that combine birefringence with TIR in such a way that light of a particular polarization is totally reflected while light of the orthogonal polarization is at least partly transmitted. Examples include the Nicol prism, Glan–Thompson prism, Glan–Foucault prism (or "Foucault prism"), and Glan–Taylor prism.

Refractometers, which measure refractive indices, often use the critical angle.

Rain sensors for automatic windscreen/windshield wipers have been implemented using the principle that total internal reflection will guide an infrared beam from a source to a detector if the outer surface of the windshield is dry, but any water drops on the surface will divert some of the light.

Edge-lit LED panels, used (e.g.) for backlighting of LCD computer monitors, exploit TIR to confine the LED light to the acrylic glass pane, except that some of the light is scattered by etchings on one side of the pane, giving an approximately uniform luminous emittance.

Total internal reflection microscopy (TIRM) uses the evanescent wave to illuminate small objects close to the reflecting interface. The consequent scattering of the evanescent wave (a form of frustrated TIR), makes the objects appear bright when viewed from the "external" side. In the total internal reflection fluorescence microscope (TIRFM), instead of relying on simple scattering, we choose an evanescent wavelength short enough to cause fluorescence. The high sensitivity of the illumination to the distance from the interface allows measurement of extremely small displacements and forces.

A beam-splitter cube uses frustrated TIR to divide the power of the incoming beam between the transmitted and reflected beams.

Optical modulation can be accomplished by means of frustrated TIR with a variable gap. As the transmission coefficient is highly sensitive to the gap width (the function being approximately exponential until the gap is almost closed), this technique can achieve a large dynamic range.

Optical fingerprinting devices have used frustrated TIR to record images of persons' fingerprints without the use of ink.

Gait analysis can be performed by using frustrated TIR with a high-speed camera, to capture and analyze footprints.

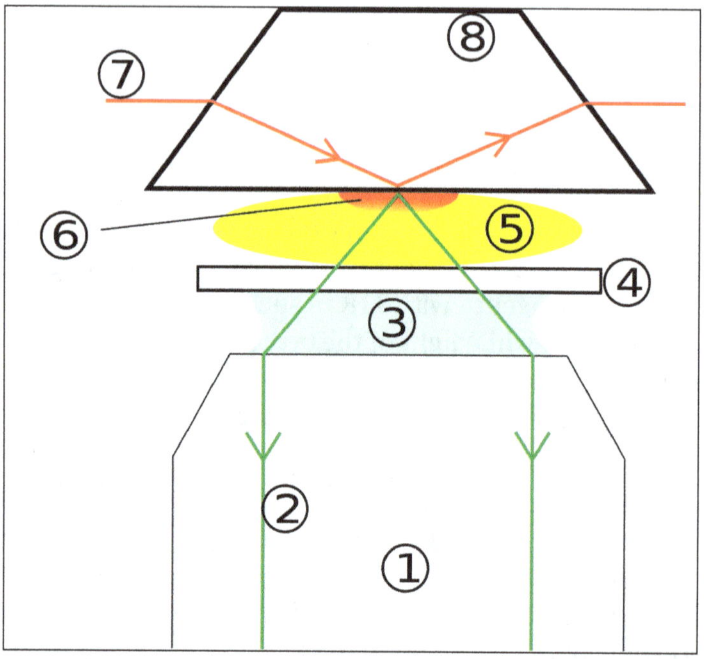

Operation of a "trans-geometry" TIR fluorescence microscope: (1) objective, (2) emission beam, (3) immersion oil, (4) cover slip, (5) specimen, (6) evanescent wave range, (7) excitation beam, (8) quartz prism.

A gonioscope, used in optometry and ophthalmology for the diagnosis of glaucoma, suppresses TIR in order to look into the angle between the iris and the cornea. This view is usually blocked by TIR at the cornea-air interface. The gonioscope replaces the air with a higher-index medium, allowing transmission at oblique incidence, typically followed by reflection in a "mirror", which itself may be implemented using TIR.

Optical Aberrations

In optics, aberration is a property of optical systems such as lenses that causes light to be spread out over some region of space rather than focused to a point. Aberrations cause the image formed by a lens to be blurred or distorted, with the nature of the distortion depending on the type of aberration. Aberration can be defined as a departure of the performance of an optical system from the predictions of paraxial optics. In an imaging system, it occurs when light from one point of an object does not converge into (or does not diverge from) a single point after transmission through the system. Aberrations occur because the simple paraxial theory is not a completely accurate model of the effect of an optical system on light, rather than due to flaws in the optical elements.

An image-forming optical system with aberration will produce an image which is not sharp. Makers of optical instruments need to correct optical systems to compensate for aberration.

Aberration can be analyzed with the techniques of geometrical optics. The articles on reflection, refraction and caustics discuss the general features of reflected and refracted rays.

References

- Geometrical-optics: brilliant.org, Retrieved 26 March, 2019
- "Comparison of Optical Aberrations". Edmund Optics. Archived from the original on Dec 6, 2011. Retrieved March 26, 2012
- Reflectionintro, lightandcolor, primer, microscope-resource: olympus-lifescience.com, Retrieved 26 March, 2019
- K.Y. Bliokh and A. Aiello (January 2013), "Goos–Hänchen and Imbert–Fedorov beam shifts: An overview", Journal of Optics, 15 (1): 014001, arXiv:1210.8236v2, doi:10.1088/2040-8978/15/1/014001
- The-Law-of-Reflection: physicsclassroom.com, Retrieved 27 April, 2019
- Geometrical-optics: brilliant.org, Retrieved 28 May, 2019
- Refraction-of-light-1455601891-1, general-knowledge: jagranjosh.com, Retrieved 29 June, 2019
- Lectures, node, teaching: ph.utexas.edu, Retrieved 30 July, 2019

3
Wave Optics

Wave optics is the branch of optics which deals with the study of interference, diffraction and polarization. Huygen's principle, Doppler effect, wavefront and wave theory of light are some of its aspects. The topics elaborated in this chapter will help in gaining a better perspective about these aspects related to wave optics.

Wave optics, also referred to as physical optics, is the branch of optics that enables important phenomena such as interference, diffraction, and polarization to be understood.

The study of these phenomena is essential for the creation of devices and concepts such as holograms, interferometers, gratings, thin-film interference, polarizers, coatings for anti-reflection (AR) and high reflection (HR), quarter-wave plates, and laser beam divergence in the near and far field.

Wave optics treats light as a series of propagating electric and magnetic field oscillations.

The phenomena influenced by wave optics are briefly explained below:

- Interference – It occurs when two or more light waves pass via the same region and add to or subtract from each other.
- Diffraction - It occurs when light waves pass via small openings or around obstacles and scatter.
- Polarization - It occurs because of the transverse nature of the electric field vibration in propagating electromagnetic wave.

Scientist Christian Huygens formulated a process for propagating waves from one position to another, thereby determining the shapes of the developing wave fronts.

Applications

The key applications that use wave optics include:

- Enhancement of microscopy.

- LCDs.
- Spectrometers.
- Michelson interferometers.

Coherent and Incoherent Addition of Waves

Suppose there is a surface of the water and you take a needle and touch the surface of the water. What will happen? Yes, ripples are formed. Now if you take two needles and you touch the surface of the water with the needles. What do you think will happen?

You will see a pattern. That pattern is the interference pattern. When you touch both the needles at the surface of the water at the same time, both the needles are in the same phase. Needle 1 will produce a wave. Also, needle 2 will produce its own ripples and they will intersect with waves of the first needle.

Now, if both the needles are moving with the same velocity, the wave formed here are coherent. If the velocity of a 1st needle and 2nd needle are not steady they won't intersect. This is because one is at a steady speed and other is at variable speed.

Browse more Topics under Wave Optics

- Diffraction.
- Huygens principle.
- Interference of light waves and Young's experiment.
- Polarisation.
- Refraction and reflection of plane waves using Huygens principle.

Coherent Waves

If the potential difference between two waves is zero or is constant w.r.t time, then the two ways are said to be coherent.

Non-coherent Waves

The waves are non-coherent if the potential difference between the two ways keeps on changing. Lightbulb, study lamp are the examples of the coherent waves. They emit waves at random potential difference.

Explanation

Now let us consider there are two needles say S_1 and S_2 moving up and down on the surface of the water and are pointing at point P. So the path difference here is given as $S_1P - S_2P$. Now the displacement by two needles and S_1S_2 are:

$$y_1 = A \cos wt$$

$$y_2 = A \cos wt$$

So the resultant displacement at point P is, $y = y_1 + y_2$. When we substitute the value of y_1 and y_2 we write,

$$y = A \cos wt + A \cos wt$$

$$y = 2A \cos wt$$

Now, we know the intensity is proportional to the square of the amplitude waves.

$$I_0 \propto A^2$$

Where I_0 is the initial intensity and A^2 is the amplitude of the wave. From equation 3, we say that A = 2A. So,

$$I_0 \propto (2A)^2 \text{ or } I_0 \propto 4A^2$$

$$I = 4 I_0$$

Now, if two needles that are S_1 are S_2 are in the same phase, the potential difference is,

$$S_1P - S_2P = n\lambda$$

Where n = 0, 1, 2,3 .... and λ = the wavelength of the wave. If the two needles S1 and S2 are vibrating at its destructive interference then, the potential difference is:

$$S_1P - S_2P = (n + 1/2)\lambda$$

Now if the potential difference of the waves is Φ then,

$$y_1 = \alpha \cos wt$$

$$y_2 = \alpha \cos wt$$

The individual intensity of each wave is I_0 , we get,

$$y = y_1 + y_2$$

$$= \alpha \cos wt + \alpha \cos(wt + \Phi)$$

$$y = 2 \quad \cos(\Phi/2)\cos(wt + \Phi/2)$$

Since, the intensity is $I_0 \propto A^2$

$$I_0 \propto 4\alpha^2 \cos^2(\Phi/2)$$

$$I = 4\, I_0 \cos^2(\Phi/2)$$

Well, the time-averaged value of $\cos^2(\Phi_t/2)$ is 1/2. So, the resultant intensity will be $I = 2I_0$ at all the points.

Wavefront

In physics, a wavefront of a time-varying field is the set (locus) of all points where the wave has the same phase of the sinusoid. The term is generally meaningful only for fields that, at each point, vary sinusoidally in time with a single temporal frequency (otherwise the phase is not well defined).

Wavefronts usually move with time. For waves propagating in a unidimensional medium, the wavefronts are usually single points; they are curves in a two dimensional medium, and surfaces in a three-dimensional one.

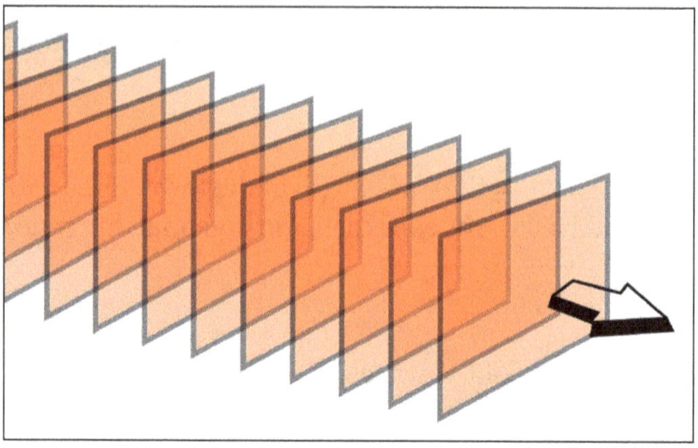

The wavefronts of a plane wave are planes.

Simple Wavefronts and Propagation

Optical systems can be described with Maxwell's equations, and linear propagating waves such as sound or electron beams have similar wave equations. However, given the above simplifications, Huygens' principle provides a quick method to predict the propagation of a wavefront through, for example, free space. The construction is as follows: Let every point on the wavefront be considered a new point source. By calculating the total effect from every point source, the resulting field at new points can be computed. Computational algorithms are often based on this approach. Specific cases for simple wavefronts can be computed directly. For example, a spherical wavefront will remain spherical as the energy of the wave is carried away equally in all directions. Such directions of energy flow, which are always perpendicular to the wavefront, are called rays creating multiple wavefronts.

The simplest form of a wavefront is the plane wave, where the rays are parallel to one another. The light from this type of wave is referred to as collimated light. The plane wavefront is a good model for a surface-section of a very large spherical wavefront; for instance, sunlight strikes the earth with a spherical wavefront that has a radius of about 150 million kilometers (1 AU). For many purposes, such a wavefront can be considered planar over distances of the diameter of Earth.

Wavefronts travel with the speed of light in all directions in an isotropic medium.

Wavefront Aberrations

Methods utilizing wavefront measurements or predictions can be considered an advanced approach to lens optics, where a single focal distance may not exist due to lens thickness or imperfections. Note also that for manufacturing reasons, a perfect lens has a spherical (or toroidal) surface shape though, theoretically, the ideal surface would be *aspheric*. Shortcomings such as these in an optical system cause what are called optical aberrations. The best-known aberrations include spherical aberration and coma.

However there may be more complex sources of aberrations such as in a large telescope due to spatial variations in the index of refraction of the atmosphere. The deviation of a wavefront in an optical system from a desired perfect planar wavefront is called the *wavefront aberration*. Wavefront aberrations are usually described as either a sampled image or a collection of two-dimensional polynomial terms. Minimization of these aberrations is considered desirable for many applications in optical systems.

Wavefront Sensor and Reconstruction Techniques

A wavefront sensor is a device which measures the wavefront aberration in a coherent signal to describe the optical quality or lack thereof in an optical system. A very common method is to use a Shack–Hartmann lenslet array. There are many applications that include adaptive optics, optical metrology and even the measurement of the aberrations in the eye itself. In this approach, a weak laser source is directed into the eye and the reflection off the retina is sampled and processed.

Alternative wavefront sensing techniques to the Shack–Hartmann system are emerging. Mathematical techniques like phase imaging or curvature sensing are also capable of providing wavefront estimations. These algorithms compute wavefront images from conventional brightfield images at different focal planes without the need for specialised wavefront optics. While Shack-Hartmann lenslet arrays are limited in lateral resolution to the size of the lenslet array, techniques such as these are only limited by the resolution of digital images used to compute the wavefront measurements.

Another application of software reconstruction of the phase is the control of telescopes through the use of adaptive optics. A common method is the Roddier test, also called wavefront curvature sensing. It yields good correction, but needs an already good system as a starting point.

The Wave Theory of Light

In the late 17th century, scientists were embroiled in a debate about the fundamental nature of light – whether it was a wave or a particle. Sir Issac Newton was a strong advocate of the particle nature of light. But, the Dutch physicist, Christiaan Huygens believed that light was made up of waves vibrating up and down perpendicular to the direction of the wave propagation, and therefore formulated a way of visualizing wave propagation. This became known as 'Huygens' Principle'.

The wave theory of light proposed by Christian Huygens has stood the test of time and today, it is considered the backbones of optics.

Light always piqued the curiosity of thinkers and scientists. But it wasn't until the late 17th century that scientists began to comprehend the properties of light. Sir Issac

Newton proposed that light was made of tiny particles known as the photons while Christian Huygens believed that light was made of waves propagating perpendicular to the direction of its movement.

In 1678, Huygen's proposed that every point that a luminous disturbance meets turns into a source of the spherical wave itself. The sum of the secondary waves, which are the result of the disturbance, determines what form the new wave will take. This theory of light is known as the 'Huygens' Principle'.

Using the above-stated principle, Huygen's was successful in deriving the laws of reflection and refraction of light. He was also successful in explaining the linear and spherical propagation of light using this theory. However, he wasn't able to explain the diffraction effects of light. Later, in 1803, the experiment conducted by Thomas Young on the interference of light proved Huygen's wave theory of light to be correct. Later in 1815, Fresnel provided mathematical equations for Young's experiment.

Max Planck proposed that light is made of finite packets of energy known as a light quantum and it depends on the frequency and velocity of light. Later, in 1905, Einstein proposed that light possessed the characteristics of both particle and wave. He suggested that light is made of small particles called photons. Quantum mechanics gave proof of the dual nature of light.

Light Wave Theory

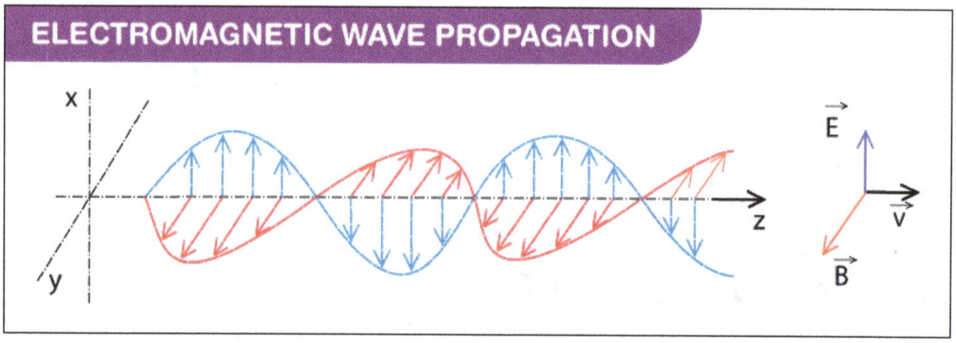

Most of the time, light behaves as a wave and it is categorised as one of the electromagnetic waves because it is made of both electric and magnetic fields. Electromagnetic fields perpendicularly oscillate to the direction of wave travel and are perpendicular to each other. As a result of which, they are known as a transverse wave. A few characteristics of light are as follows:

- While dealing with light waves, we deal with the sine waveform. The period of the waveform is one full 0 to 360 degree sweep.
- Light waves have two important characteristics known as wavelength and frequency.

- The distance between the peaks of the wave is known as the wavelength. In case of a light wave, the wavelengths are in the order of nanometers.
- Frequency is the number of waves that will cross past a point in a second.

The relationship between wavelength and frequency is given by the equation: $f = \frac{1}{T}$

- The speed of light in a vacuum is a universal constant which is 3×10^8 m/s.
- As proposed by Einstein, light is made of tiny packets of energy known as photons. The formula devised by Planck determines the energy of a photon and it also shows that the energy is directly proportional to the frequency of the light.

  $E = hf$

where h is the Planck's constant 6.63×10^{-34} Joule-Second

Huygen's Principle

Some phenomena require analysis and explanations based on the wave characteristics of light. This is particularly true when the wavelength is not negligible compared to the dimensions of an optical device, such as a slit in the case of diffraction. Huygens's principle is an indispensable tool for this analysis.

In figure below how a transverse wave looks as viewed from above and from the side. A light wave can be imagined to propagate like this, although we do not actually see it wiggling through space. From above, we view the wave fronts (or wave crests) as if we were looking down on ocean waves. The side view would be a graph of the electric or magnetic field. The view from above is perhaps more useful in developing concepts about wave optics.

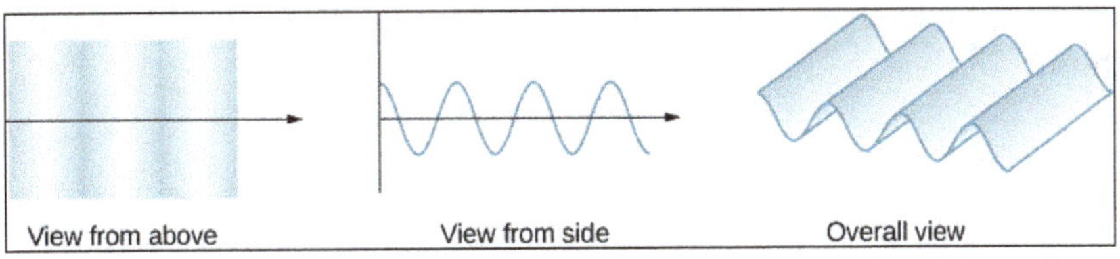

A transverse wave, such as an electromagnetic light wave, as viewed from above and from the side. The direction of propagation is perpendicular to the wave fronts (or wave crests) and is represented by a ray.

The Dutch scientist Christiaan Huygens developed a useful technique for determining in detail how and where waves propagate. Starting from some known position, Huygens's principle states that every point on a wave front is a source of wavelets that spread out in the forward direction at the same speed as the wave itself. The new wave front is tangent to all of the wavelets.

In figure below how Huygens's principle is applied. A wave front is the long edge that moves, for example, with the crest or the trough. Each point on the wave front emits a semicircular wave that moves at the propagation speed v. We can draw these wavelets at a time t later, so that they have moved a distance $s = vt$. The new wave front is a plane tangent to the wavelets and is where we would expect the wave to be a time t later. Huygens's principle works for all types of waves, including water waves, sound waves, and light waves. It is useful not only in describing how light waves propagate but also in explaining the laws of reflection and refraction. In addition, we will see that Huygens's principle tells us how and where light rays interfere.

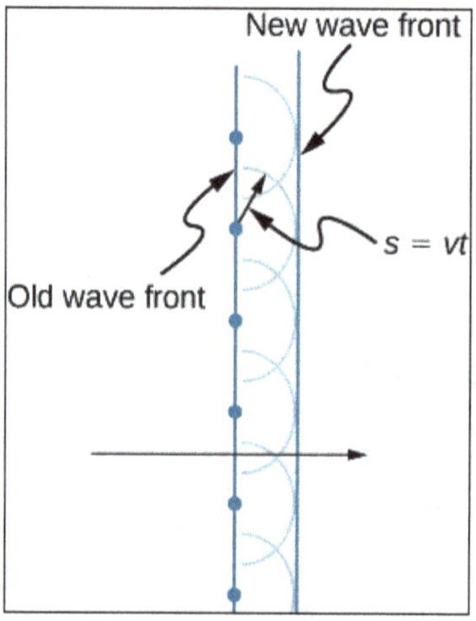

Huygens's principle applied to a straight wave front. Each point on the wave front emits a semicircular wavelet that moves a distance $s = vt$. The new wave front is a line tangent to the wavelets.

Reflection

In figure below how a mirror reflects an incoming wave at an angle equal to the incident angle, verifying the law of reflection. As the wave front strikes the mirror, wavelets are first emitted from the left part of the mirror and then from the right. The wavelets closer to the left have had time to travel farther, producing a wave front traveling in the direction shown.

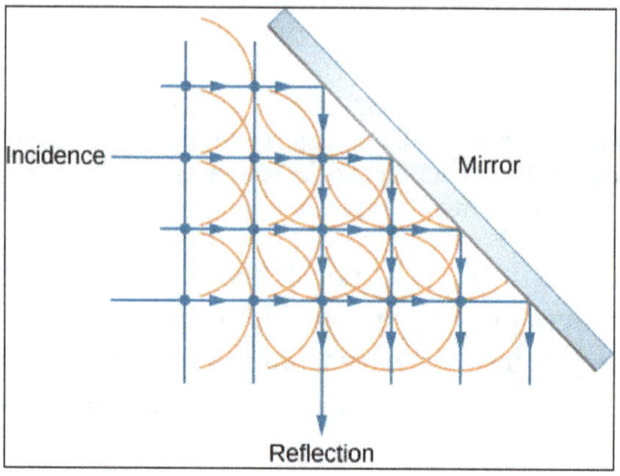

Huygens's principle applied to a plane wave front striking a mirror. The wavelets shown were emitted as each point on the wave front struck the mirror. The tangent to these wavelets shows that the new wave front has been reflected at an angle equal to the incident angle. The direction of propagation is perpendicular to the wave front, as shown by the downward-pointing arrows.

Refraction

The law of refraction can be explained by applying Huygens's principle to a wave front passing from one medium to another. Each wavelet in the figure was emitted when the wave front crossed the interface between the media. Since the speed of light is smaller in the second medium, the waves do not travel as far in a given time, and the new wave front changes direction as shown. This explains why a ray changes direction to become closer to the perpendicular when light slows down. Snell's law can be derived from the geometry in figure.

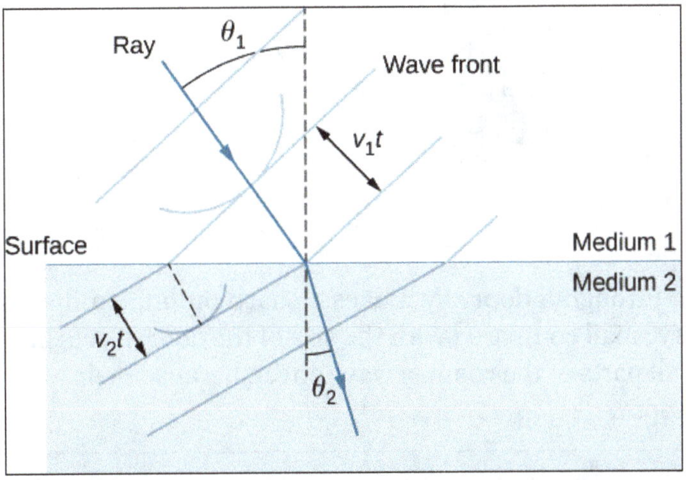

Huygens's principle applied to a plane wave front traveling from one medium to another, where its speed is less. The ray bends toward the perpendicular, since the wavelets have a lower speed in the second medium.

This applet by Walter Fendt shows an animation of reflection and refraction using Huygens's wavelets while you control the parameters.

Diffraction

What happens when a wave passes through an opening, such as light shining through an open door into a dark room? For light, we observe a sharp shadow of the doorway on the floor of the room, and no visible light bends around corners into other parts of the room. When sound passes through a door, we hear it everywhere in the room and thus observe that sound spreads out when passing through such an opening. What is the difference between the behavior of sound waves and light waves in this case? The answer is that light has very short wavelengths and acts like a ray. Sound has wavelengths on the order of the size of the door and bends around corners (for frequency of 1000 Hz, about three times smaller than the width of the doorway).

$$\lambda = \frac{c}{f} = \frac{330\ m/s}{1000\ s^{-1}} = 0.33\ m,$$

If we pass light through smaller openings such as slits, we can use Huygens's principle to see that light bends as sound does. The bending of a wave around the edges of an opening or an obstacle is called diffraction. Diffraction is a wave characteristic and occurs for all types of waves. If diffraction is observed for some phenomenon, it is evidence that the phenomenon is a wave. Thus, the horizontal diffraction of the laser beam after it passes through the slits in figure is evidence that light is a wave.

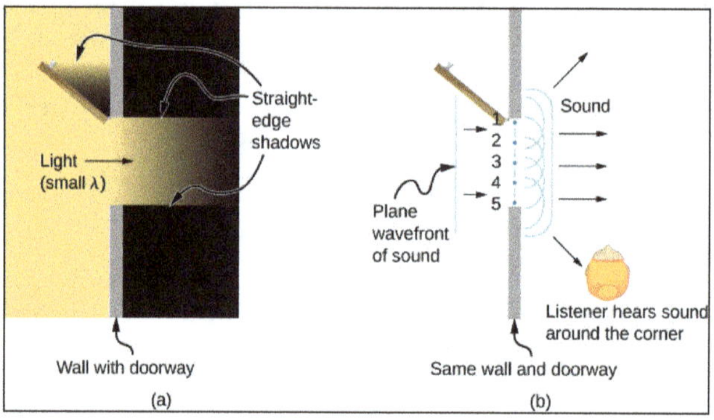

(a) Light passing through a doorway makes a sharp outline on the floor. Since light's wavelength is very small compared with the size of the door, it acts like a ray. (b) Sound waves bend into all parts of the room, a wave effect, because their wavelength is similar to the size of the door.

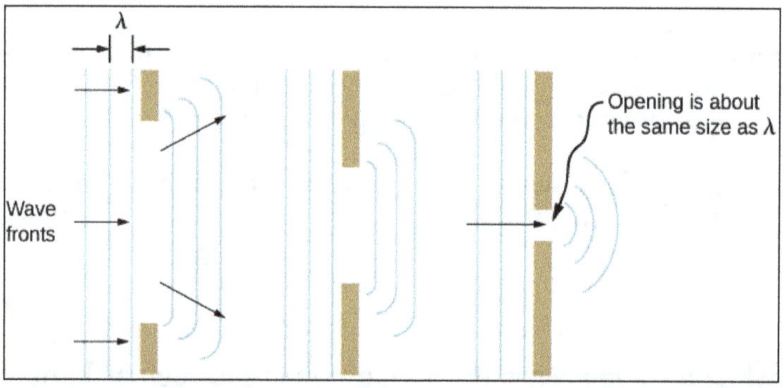

Huygens's principle applied to a plane wave front striking an opening. The edges of the wave front bend after passing through the opening, a process called diffraction. The amount of bending is more extreme for a small opening, consistent with the fact that wave characteristics are most noticeable for interactions with objects about the same size as the wavelength.

Interference of Light

Interference of Light Waves is defined as the modification in the distribution of light energy when two or more waves superimpose each other. For Interference the waves emitted by sources should be with zero phase difference or no phase difference. These sources should emit continuous waves of same wave length and same time period. These sources should be very close to each other. The sources which emit light waves should be coherent sources. If the waves are coherent, the interference pattern is observable and is stable. If the waves are incoherent, the pattern is not visible.

The wave theory of light was proposed by Christiaan Huygens. Huygens stated that light is made up of waves that vibrates up and down and is perpendicular to the direction of light. But this concept was not readily accepted and was in contradiction to the particle theory or the corpuscular theory of light proposed by Sir Isaac Newton. Isaac Newton described that light is made up of tiny particles. In the beginning, wave theory of light was not accepted.

Later, another scientist named Thomas Young performed the famous Interference experiment, which is called the Young's Double Slit Experiment (YDSE). This experiment proved the wave nature of light. After this, many other experiments were carried out which included the Interference, diffraction and polarization patterns. These experiments were only explained by considering the wave nature of light. Thus wave theory came into existence. Corpuscular theory could not explain the various optical phenomena like interference, diffraction, polarization, dispersion and many others.

Types of Interference

There are two types of Interference:

Constructive Interference

When crest of one wave falls on the crest of other wave, the resultant amplitude will be maximum. The case in which the resultant amplitude is maximum is called constructive Interference. Here the two interfering waves have a displacement which is in the same direction.

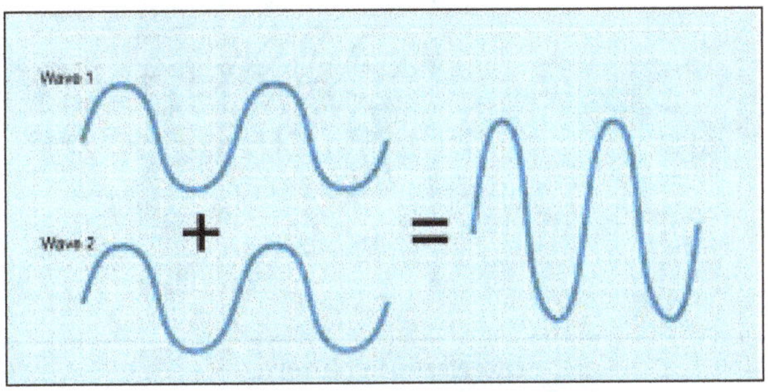

Constructive Interference.

Consider two waves which travels in same direction with same frequency. Now the amplitude of the waves gets add up when the light waves are at same place and time. The new wave looks similar to the original waves, but the amplitude of the wave is higher. In constructive interference, the waves are in phase.

Destructive Interference

When crest of one wave falls on the trough of other wave, the resultant amplitude will be minimum. The case in which resultant amplitude is minimum is called Destructive Interference. Consider the figure given below. Here the first wave is up and the second wave is down. So when they add together the resultant one becomes zero. Thus in destructive interference, the sum of the wave can either be less than the original waves and or it can be zero. Here the two waves will be out of phase.

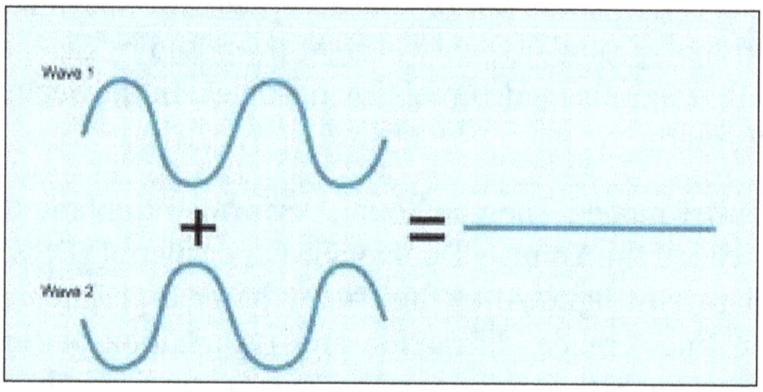

Destructive Interference.

Interference Pattern

Constructive Interference results in bright bands and destructive interference results in dark bands. These are called Fringes. This formation of bright and dark bands on the screen results in forming an interference pattern. Consider the picture shown below. The two slits present in it act as two coherent sources of light. We can see the pattern of light and dark bands here.

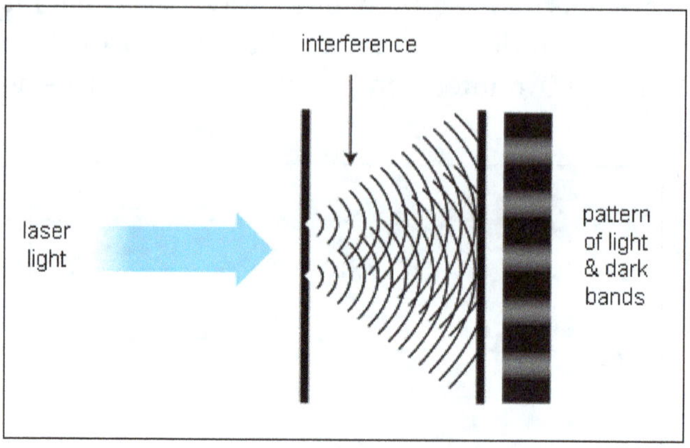

Interference pattern.

Intensity of Light at the Point of Interference

We know that intensity of any light wave is directly proportional to the square of its amplitude. In case of coherent sources, let's consider the factor of frequency to be a constant. Thus the intensity for the wave is considered as KA^2.

$I = KA^2$, where K is the constant which depends on what medium the wave is in.

Consider the resultant amplitude as 'R' at the point of interference. Now the resultant intensity at this point can be written as:

$$I_R = KR^2$$
$$= K\left[A_1^2 + A_2^2 + 2A_1A_2 \ \text{Cos}\,\phi\right]$$
$$= KA_1^2 + KA_2^2 + 2KA_1A_2\text{Cos}\,\phi$$

Substitute $KA_1^2 = I_1$ and $KA_2^2 = I_2$

Thus $I_R = I_1 + I_2 + 2\sqrt{(I_1I_2\text{Cos}\,\phi)}$

So the resultant intensity of two waves after the interference phenomenon is:

$$I_R = I_1 + I_2 + 2\sqrt{(I_1I_2\text{Cos}\,\phi)}$$

Here ϕ is the phase difference between two waves.

In case of constructive interference, the value of $\phi = 0$ and so $\text{Cos}\,\phi = 1$.Then $I_R = I_1 + I_2 + 2(\sqrt{I_1I_2} = \left(\sqrt{I_1} + \sqrt{I_2}\right)^2$ where the waves are superposed in same phase. Here the resultant intensity is maximum. For destructive interference, the waves superpose in opposite direction. Then $\text{Cos}\,\phi = -1$ Thus the resultant intensity is minimum. So, $I_R = I_1 + I_2 - 2\sqrt{(I_1I_2} = \left(\sqrt{I_1} - \sqrt{I_2}\right)^2$.

Interference of Equal Intensity Waves

As mentioned earlier, the resultant intensity of waves at the point of interference

$I_R = (I_1 + I_2 + 2\sqrt{(I_1I_2\text{Cos}\,\phi)}$. Let us consider that the two waves have same and equal intensity, which means $I_1 = I_2 = I_0$.

Thus $I_R = 2I_0 + 2I_0\ \text{Cos}\,\phi$

$$= 2I_0(1 + \text{Cos}\,\phi)$$

$$1 + \text{Cos}\,\phi = 2\cos^2\phi/2 \ (\text{Cos}\,\phi \text{ in half angle form})$$

Then $I_R = 4I_0 \cos^2 \phi/2$.

Hence for constructive interference, intensity will be maximum, $I_R = \left(\sqrt{I_1} + \sqrt{I_2}\right)^2 = 4\ I_0$.

For destructive interference, the intensity is minimum and $I_R = \left(\sqrt{I_1} + \sqrt{I_2}\right)^2 = 0$.

Path Difference for Constructive and Destructive Interference

In constructive interference, the phase difference is considered as 2nЛ, where n being an integer. Now to find the path difference between the waves,

$$\Delta = /2\pi * \phi$$

$$= /2\pi * 2n\pi$$

$$\Delta = n\lambda (n = 0, 1, 2\ldots\ldots)$$

Hence, we can say that two waves interfere constructively, when their path difference $\Delta = \lambda, 2\lambda \ldots .\ n\lambda ..$

Consider two light sources S_1 and S_2, which are coherent. Assume that the light sources are switched on at the same time. Consider a point P where the two light waves emitted from the coherent sources are superposed. Now the point P is located at a distance a1 from source S1 and a distance a2 from source a2. If the two sources are in same phase, then the path difference, $\Delta = a_2 - a_1$. If the path difference at point P with respect to the two sources is multiple of λ, then the intensity at point p is maximum and will be equal to 4Io, if the two waves are of equal intensity. Otherwise it is shown as $\left(\sqrt{I_1} + \sqrt{I_2}\right)^2$. In case of destructive interference, we know that $\text{Cos}\ \phi = -1$ and it happens when the phase difference, $\phi = (2n + 1)\pi$ which corresponds to opposite phase. To calculate the path difference between the waves in destructive interference,

$$\phi = /2\pi * \phi$$

$$= \lambda / 2\pi * (2n+1)\pi$$

$$= (2n + 1)\ \lambda / 2$$

Hence, we can say that two waves interfere destructively, when their path difference,

$$\Delta = \lambda / 2,\ 3\lambda / 2 \ldots .(2n - 1)\lambda / 2.(n = 1, 2\ldots\ldots)$$

For destructive interference, the waves superpose in opposite direction. Then $\text{Cos}\ \phi = -1$ Thus the resultant intensity is minimum. So, $I_R = I_1 + I_2 - 2\sqrt{(I_1 I_2} = \sqrt{(I_1 - I_2)^2}$.

Young's Double Slit Experiment

As mentioned earlier, Thomas young performed the famous interference experiment which is the Ydse. It was by this experiment; the wave theory of light came into existence.

In this experiment, Thomas Young used one single light source and it was the passed through two slits to obtain two coherent sources. Thus each slit act as a light source.

Consider two slits S1 and S2 on a screen which is kept parallel and very close to each other. These slits are illuminated by other narrow slit S. Light spreads out fromS1 and S2 ad falls on a screen. When both slits are open, there is a formation of interference fringes.

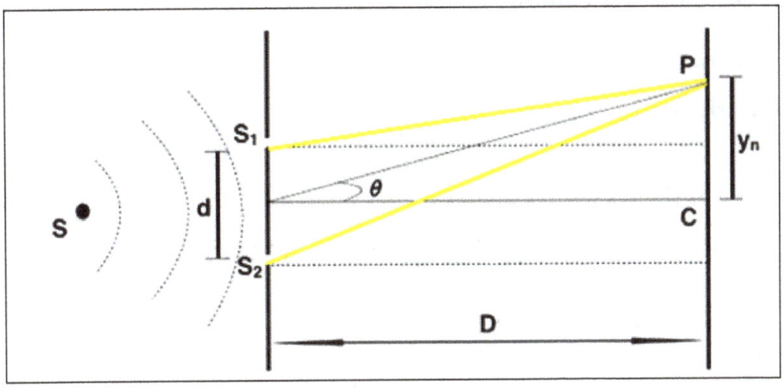

Young's double slit experiment.

We know that $I = I_1 + I_2 + 2\sqrt{(I_1\ I_2\ \text{Cos}\ \phi)}$.

Condition for bright fringes/maxima, $\phi = 2n\Pi$ or $\Delta = n\lambda$, path difference

Where $n = 0, 1, 2$. Φ is the phase difference.

Condition for dark fringes/minima, $\phi = (2n-1)\Pi$ or $\Delta = (n-1)/2\lambda$, path difference.

Where $n = 1, 2, 3$.

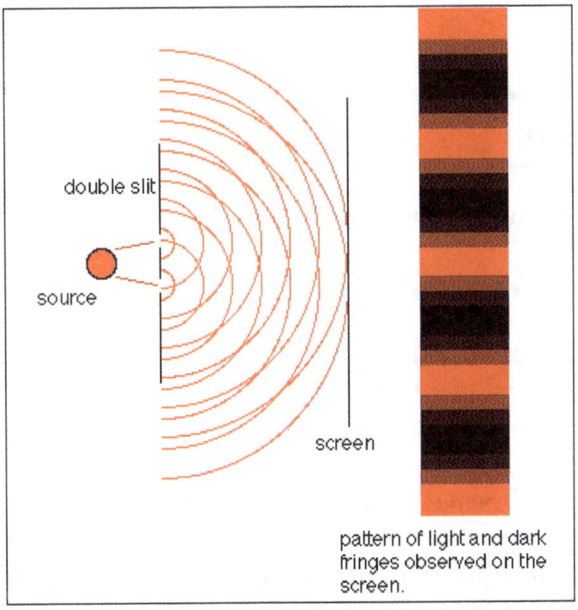

Dark and bright fringes.

Maxima and Minima

Path difference between two waves,

$$\Delta = S_1P - S_2P = d \sin \theta$$

We know that D >> d, so the angle θ is small,

$$\sin\theta = Y_n / D$$
$$\Delta = d \sin \theta = d\left(Y_n / D\right) \Rightarrow Y_n = \Delta\ D / d$$

For nth maxima, $\Delta = n\lambda$

$Y_n = n\lambda D/d$, where n = 0, 1, 2, 3. This denotes bright fringe.

$$n = 0, Y_n = 0$$
$$n = 1, Y_n = \lambda D/d$$
$$n = 2, Y_n = 2\lambda D/d$$

1 ane,

For nth minima $\Delta = \left(n\text{-}1/2\right)\lambda$

$Y_n = \left(n - ½\right)\lambda D/d$ where n = 1, 2, 3. This denotes dark fringe,

$$n = 1,\ Y_n = \lambda D/2d$$
$$n = 2,\ Y_n = 3\lambda D/2d$$

Polarisation

The process of transforming unpolarized light into polarized light is known as polarization.

Polarization by use of a Polaroid Filter

The most common method of polarization involves the use of a Polaroid filter. Polaroid filters are made of a special material that is capable of blocking one of the two planes of vibration of an electromagnetic wave. (Remember, the notion of two planes or directions of vibration is merely a simplification that helps us to visualize the wavelike nature of the electromagnetic wave.) In this sense, a Polaroid serves as a device that filters out one-half of the vibrations upon transmission of the light through the filter. When unpolarized light is transmitted through a Polaroid filter, it emerges with one-half the intensity and with vibrations in a single plane; it emerges as polarized light.

A Polaroid filter is able to polarize light because of the chemical composition of the filter material. The filter can be thought of as having long-chain molecules that are aligned within the filter in the same direction. During the fabrication of the filter, the long-chain molecules are stretched across the filter so that each molecule is (as much as possible) aligned in say the vertical direction. As unpolarized light strikes the filter, the portion of the waves vibrating in the vertical direction are absorbed by the filter. The general rule is that the electromagnetic vibrations that are in a direction parallel to the alignment of the molecules are absorbed.

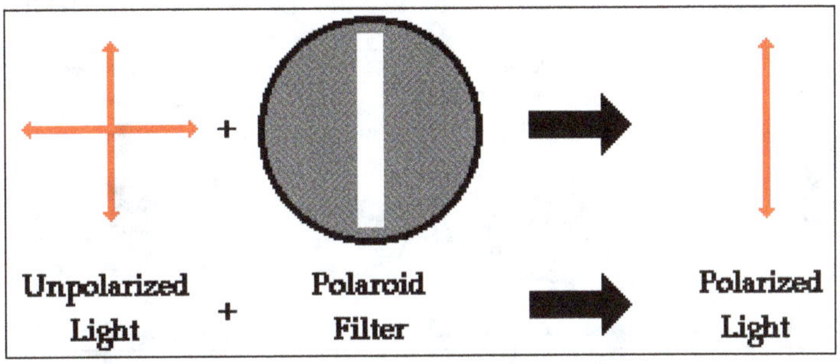

The alignment of these molecules gives the filter a polarization axis. This polarization axis extends across the length of the filter and only allows vibrations of the electromagnetic wave that are parallel to the axis to pass through. Any vibrations that are perpendicular to the polarization axis are blocked by the filter. Thus, a Polaroid filter with its long-chain molecules aligned horizontally will have a polarization axis aligned vertically. Such a filter will block all horizontal vibrations and allow the vertical vibrations to be transmitted. On the other hand, a Polaroid filter with its long-chain molecules aligned vertically will have a polarization axis aligned horizontally; this filter will block all vertical vibrations and allow the horizontal vibrations to be transmitted.

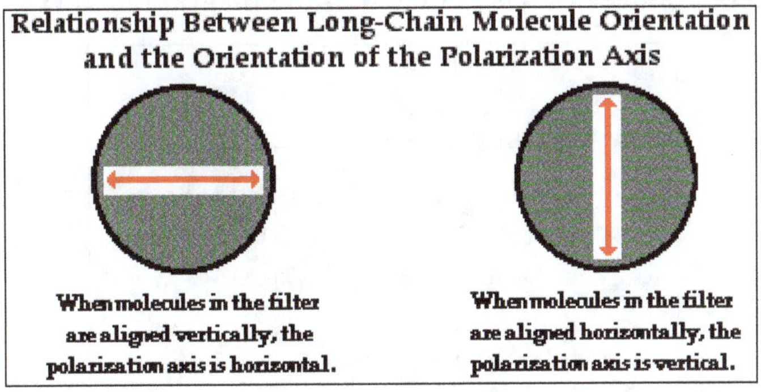

Polarization of light by use of a Polaroid filter is often demonstrated in a Physics class through a variety of demonstrations. Filters are used to look through and view objects. The filter does not distort the shape or dimensions of the object; it merely serves to produce a dimmer image of the object since one-half of the light is blocked as it passed through the filter. A pair of filters is often placed back to back in order to view objects looking through two filters. By slowly rotating the second filter, an orientation can be found in which all the light from an object is blocked and the object can no longer be seen when viewed through two filters. What happened? In this demonstration, the light was polarized upon passage through the first filter; perhaps only vertical vibrations were able to pass through. These vertical vibrations were then blocked by the second filter since its polarization filter is aligned in a horizontal direction. While you are unable to see the axes on the filter, you will know when the axes are aligned perpendicular to each other because with this orientation, all light is blocked. So by use of two filters, one can completely block all of the light that is incident upon the set; this will only occur if the polarization axes are rotated such that they are perpendicular to each other.

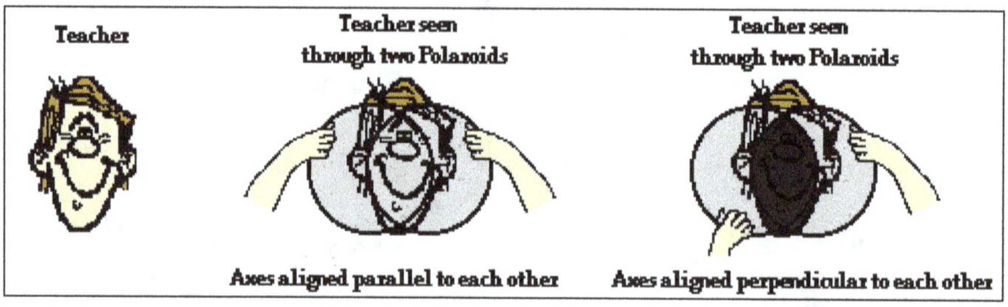

A picket-fence analogy is often used to explain how this dual-filter demonstration works. A picket fence can act as a polarizer by transforming an unpolarized wave in a rope into a wave that vibrates in a single plane. The spaces between the pickets of the fence will allow vibrations that are parallel to the spacings to pass through while blocking any vibrations that are perpendicular to the spacings. Obviously, a vertical vibration would not have the room to make it through a horizontal spacing. If two picket fences are oriented such that the pickets are both aligned vertically, then vertical vibrations will pass through both fences. On the other hand, if the pickets of the second fence are aligned horizontally, then the vertical vibrations that pass through the first fence will be blocked by the second fence. This is depicted in the diagram below.

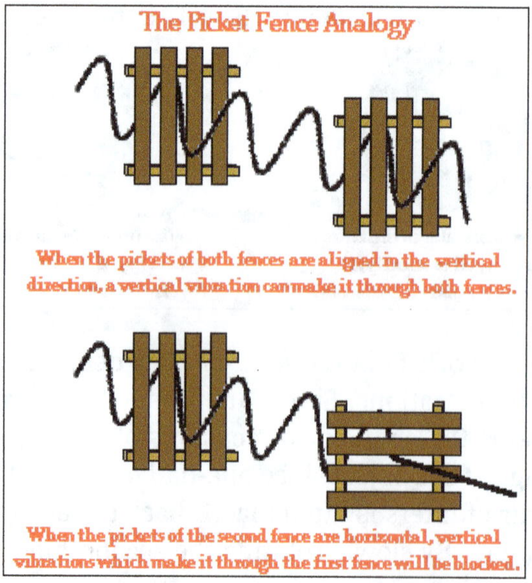

In the same manner, two Polaroid filters oriented with their polarization axes perpendicular to each other will block all the light. Now that's a pretty cool observation that could never be explained by a particle view of light.

Polarization by Reflection

Unpolarized light can also undergo polarization by reflection off of nonmetallic surfaces. The extent to which polarization occurs is dependent upon the angle at which the light approaches the surface and upon the material that the surface is made of. Metallic surfaces reflect light with a variety of vibrational directions; such reflected light is unpolarized. However, nonmetallic surfaces such as asphalt roadways, snowfields and water reflect light such that there is a large concentration of vibrations in a plane parallel to the reflecting surface. A person viewing objects by means of light reflected off of nonmetallic surfaces will often perceive a glare if the extent of polarization is large. Fishermen are familiar with this glare since it prevents them from seeing fish that lie below the water. Light reflected off a lake is partially polarized in a direction parallel to the water's surface. Fishermen know that the use of glare-reducing sunglasses with the proper polarization axis allows for the blocking of this partially polarized light. By blocking the plane-polarized light, the glare is reduced and the fisherman can more easily see fish located under the water.

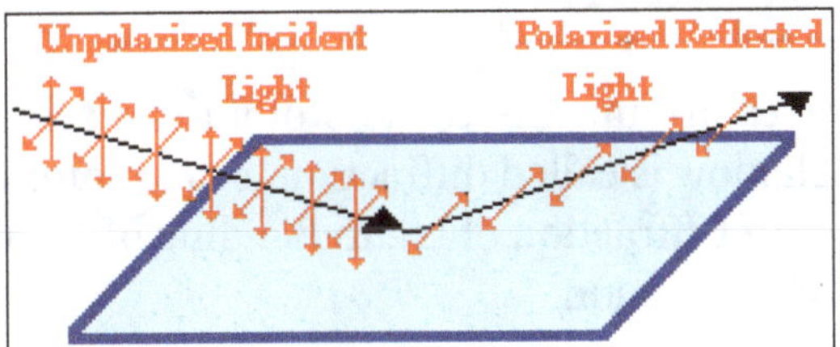

Reflection of light off of non-metallic surfaces results in some degree of polarization parallel to the surface.

Polarization by Refraction

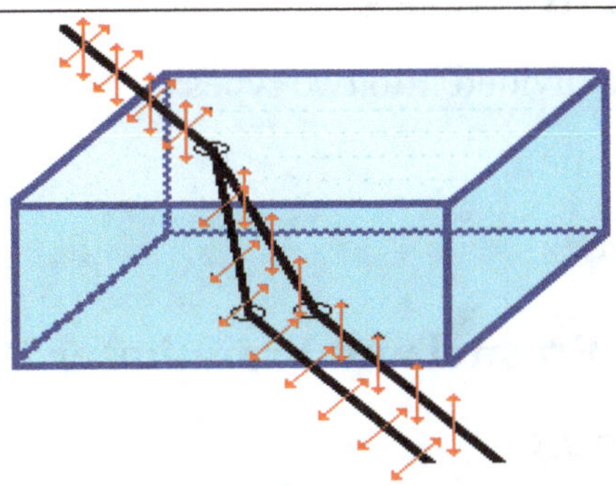

The two refracted rays passing through the Iceland Spar crystal are polarized with perpendicular orientations.

Polarization can also occur by the refraction of light. Refraction occurs when a beam of light passes from one material into another material. At the surface of the two materials, the path of the beam changes its direction. The refracted beam acquires some degree of polarization. Most often, the polarization occurs in a plane perpendicular to the surface. The polarization of refracted light is often demonstrated in a Physics class using a unique crystal that serves as a double-refracting crystal. Iceland Spar, a rather rare form of the mineral calcite, refracts incident light into two different paths. The light is split into two beams upon entering the crystal. Subsequently, if an object is viewed by looking through an Iceland Spar crystal, two images will be seen. The two images are the result of the double refraction of light. Both refracted light beams are polarized - one in a direction parallel to the surface and the other in a direction perpendicular to the surface. Since these two refracted rays are polarized with a perpendicular orientation, a polarizing filter can be used to completely block one of the images. If the polarization axis of the filter is aligned perpendicular to the plane of polarized light, the light is completely blocked by the filter; meanwhile the second image is as bright as can be. And if the filter is then turned 90-degrees in either direction, the second image reappears and the first image disappears. Now that's pretty neat observation that could never be observed if light did not exhibit any wavelike behavior.

Diffraction of Light

The bending of light waves around the corners of an obstacle and spreading of light waves into geometrical shadow is called diffraction. Fraunhofer Diffraction and Fresnel Diffraction are two Types of Diffraction of Light. Bending of Light around the corners of Window is an example of Diffraction.

Diffraction effect depends upon the size of the obstacle. Diffraction of light takes place if the size of the obstacle is comparable to the wavelength of light. Light waves are very small in wavelength,i.e, from 4×10^{-7} m to 7×10^{-7} m. If the size of opening or obstacle is near to this limit, only then we can observe the phenomenon of diffraction.

Types of Diffraction in Physics

Diffraction of light can be divided into two types:

- Fraunhofer Diffraction.
- Fresnel Diffraction.

Difference between Fresnel and Fraunhofer Diffraction

Fraunhofer Diffraction

In Fraunhofer diffraction:

- Source and the screen are far away from each other.

- Incident wavefronts on the diffracting obstacle are plane.
- Diffraction obstacle gives rise to wavefronts which are also plane.
- Plane diffracting wavefronts are converged by means of a convex lens to produce a diffraction pattern.

Fresnel Diffraction

In Fresnel diffraction:

- Source and screen are not far away from each other.
- Incident wavefronts are spherical.
- Wavefronts leaving the obstacles are also spherical.
- The convex lens is not needed to converge the spherical wavefronts.

In Young's double-slit experiment for the interference of light, the central region of the fringe system is bright. If light travels in a straight path, the central region should appear dark i.e., the shadow of the screen between the two slits. Another simple experiment can be performed by exhibiting the same effect.

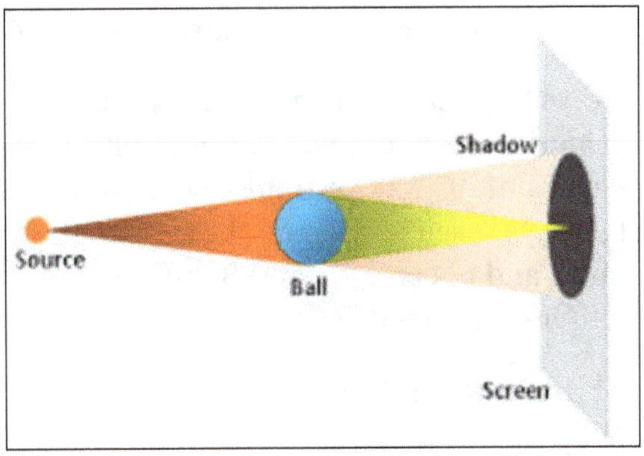

Consider that a small and smooth ball of about 3 mm in diameter is illuminated by a point source of light.The shadow of the object is received on a screen as shown in the figure. The shadow of the spherical object is not completely dark but has a bright spot at its centre. According to Huygens's principle, each point on the rim of the sphere behaves as a source of secondary wavelets which illuminate the central region of the shadow.

These two experiments clearly show that when light travels past an obstacle, it does not proceed exactly along a straight path, but bends around the obstacle. The phenomenon is found to be prominent when the wavelength of light is compared with the size of the obstacle or aperture of the slit. The diffraction of light occurs, in effect, due to the interference between rays coming from different parts of the same wavefront.

Diffraction Due to Narrow Slit

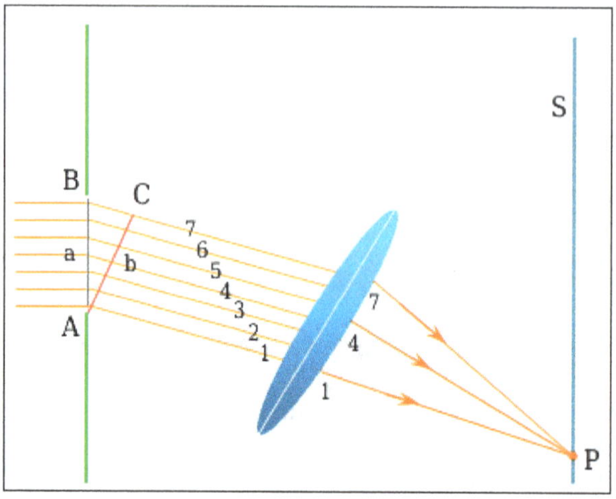

The figure shows the experimental arrangement for studying diffraction of light due to the narrow slit. The slit AB of width d is illuminated by a parallel beam of monochromatic light of wavelength λ. The screen S is placed parallel to the slit for observing the effects of the diffraction of light. A small portion of the incident wavefront passes through the narrow slit. Each point of this section of the wavefront sends out secondary wavelets to the screen. These wavelets then interfere to produce the diffraction pattern. It becomes simple to deal with rays instead of wavefronts as shown in the figure.

In this figure, only nine rays have been drawn whereas actually there are a large number of them. Let us consider rays 1 and 5 which are in phase on in the wavefront AB. When these reach the wavefront AC, ray 5 would have a path difference ab say equal to λ/2. Thus, when these two rays reach point p on the screen; they will interfere destructively. Similarly, each pair 2 and 6,3 and 7,4 and 8 differ in the path by λ/2 and will do the same. But the path difference $ab = d/2 \sin\theta$.

The equation for the first minimum is, then

$$d/2 \sin\theta = \lambda/2$$

$$\text{or } d \sin\theta = \lambda$$

In general, the conditions for different orders of minima on either side of the centre are given by:

$$d \sin\theta = m\lambda$$

$$\text{where } m = \pm\ (1,2,3,\ldots.)$$

The region between any two consecutive minima both above and below O will be bright. A narrow slit, therefore, produces a series of bright and dark regions with the first bright region at the centre of the pattern.

Single-slit Diffraction

A long slit of infinitesimal width which is illuminated by light diffracts the light into a series of circular waves and the wavefront which emerges from the slit is a cylindrical wave of uniform intensity, in accordance with Huygens–Fresnel principle.

A slit which is wider than a wavelength produces interference effects in the space downstream of the slit. These can be explained by assuming that the slit behaves as though it has a large number of point sources spaced evenly across the width of the slit. The analysis of this system is simplified if we consider light of a single wavelength. If the incident light is coherent, these sources all have the same phase. Light incident at a given point in the space downstream of the slit is made up of contributions from each of these point sources and if the relative phases of these contributions vary by 2π or more, we may expect to find minima and maxima in the diffracted light. Such phase differences are caused by differences in the path lengths over which contributing rays reach the point from the slit.

We can find the angle at which a first minimum is obtained in the diffracted light by the following reasoning. The light from a source located at the top edge of the slit interferes destructively with a source located at the middle of the slit, when the path difference between them is equal to $\lambda/2$. Similarly, the source just below the top of the slit will interfere destructively with the source located just below the middle of the slit at the same angle. We can continue this reasoning along the entire height of the slit to conclude that the condition for destructive interference for the entire slit is the same as the condition for destructive interference between two narrow slits a distance apart that is half the width of the slit. The path difference is approximately $\frac{d\sin(\theta)}{2}$ so that the minimum intensity occurs at an angle θ_{min} given by:

$$d\sin\theta_{min} = \lambda$$

where,

- d is the width of the slit,
- θ_{min} is the angle of incidence at which the minimum intensity occurs,
- λ is the wavelength of the light.

A similar argument can be used to show that if we imagine the slit to be divided into four, six, eight parts, etc., minima are obtained at angles θ_n given by:

$$d\sin\theta_n = n\lambda$$

where,

- n is an integer other than zero.

There is no such simple argument to enable us to find the maxima of the diffraction pattern. The intensity profile can be calculated using the Fraunhofer diffraction equation as:

$$I(\theta) = I_0 \operatorname{sinc}^2\left(\frac{d\pi}{\lambda}\sin\theta\right)$$

where,

- $I(\theta)$ is the intensity at a given angle,
- I_0 is the original intensity,
- The unnormalized sinc function above is given by $\operatorname{sinc}(x) = \frac{\sin x}{x}$ for $x \neq 0$, and $\operatorname{sinc}(0) = 1$

This analysis applies only to the far field, that is, at a distance much larger than the width of the slit.

From the intensity profile above, if $d \ll \lambda$, the intensity will have little dependency on θ, hence the wavefront emerging from the slit would resemble a cylindrical wave of uniform intensity; If $d \gg \lambda$, only $\theta \approx 0$ would have appreciable intensity, hence the wavefront emerging from the slit would resemble that of geometrical optics.

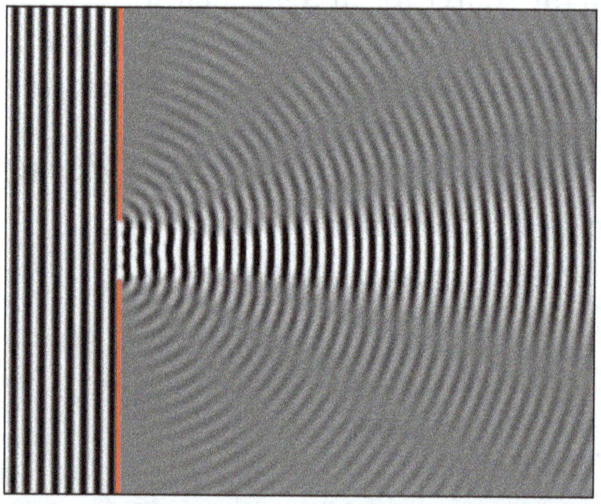

Numerical approximation of diffraction pattern from a slit of width four wavelengths with an incident plane wave. The main central beam, nulls, and phase reversals are apparent.

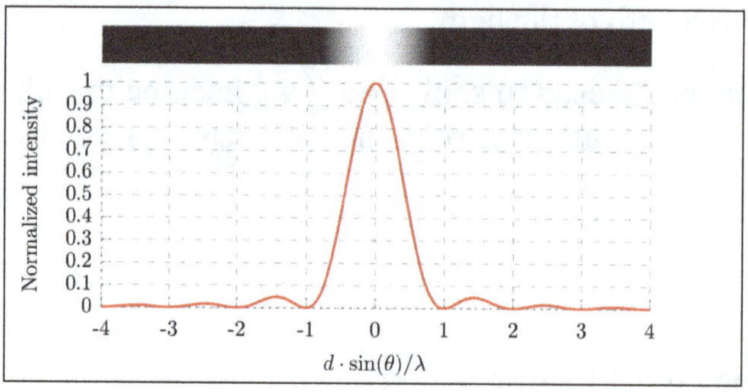

Graph and image of single-slit diffraction.

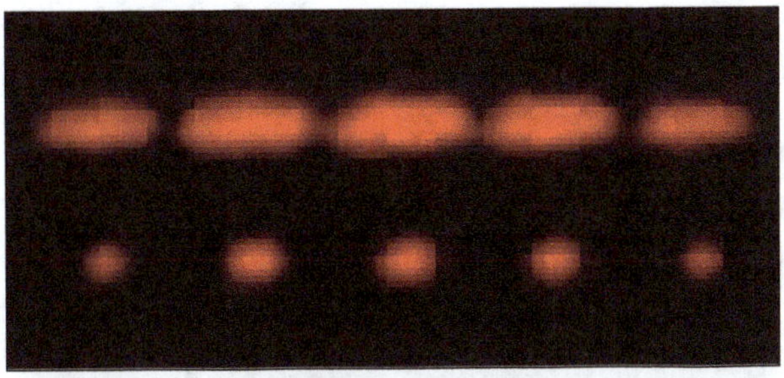

2-slit (top) and 5-slit diffraction of red laser light.

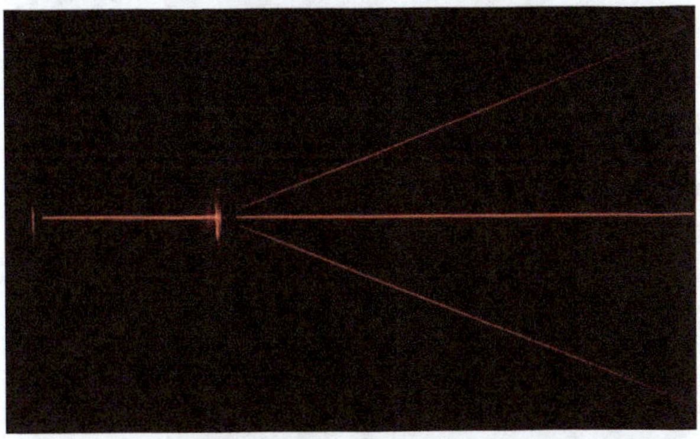

Diffraction of a red laser using a diffraction grating.

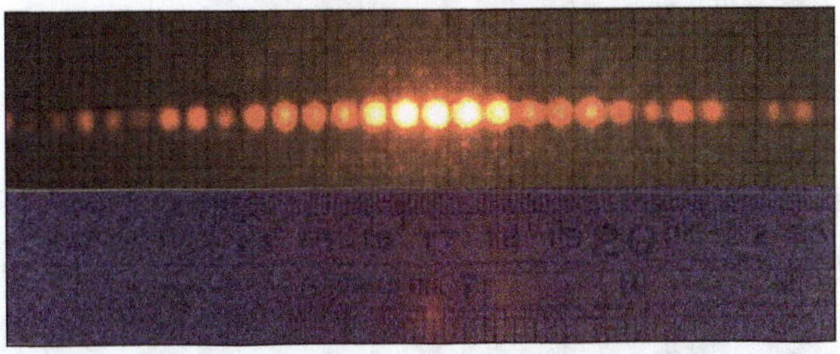

A diffraction pattern of a 633 nm laser through a grid of 150 slits.

Diffraction Grating

A diffraction grating is an optical component with a regular pattern. The form of the light diffracted by a grating depends on the structure of the elements and the number of elements present, but all gratings have intensity maxima at angles θ_m which are given by the grating equation:

$$d\left(\sin\theta_m + \sin\theta_i\right) = m\lambda.$$

where,

- θ_i is the angle at which the light is incident,
- d is the separation of grating elements,
- m is an integer which can be positive or negative.

The light diffracted by a grating is found by summing the light diffracted from each of the elements, and is essentially a convolution of diffraction and interference patterns.

The figure shows the light diffracted by 2-element and 5-element gratings where the grating spacings are the same; it can be seen that the maxima are in the same position, but the detailed structures of the intensities are different.

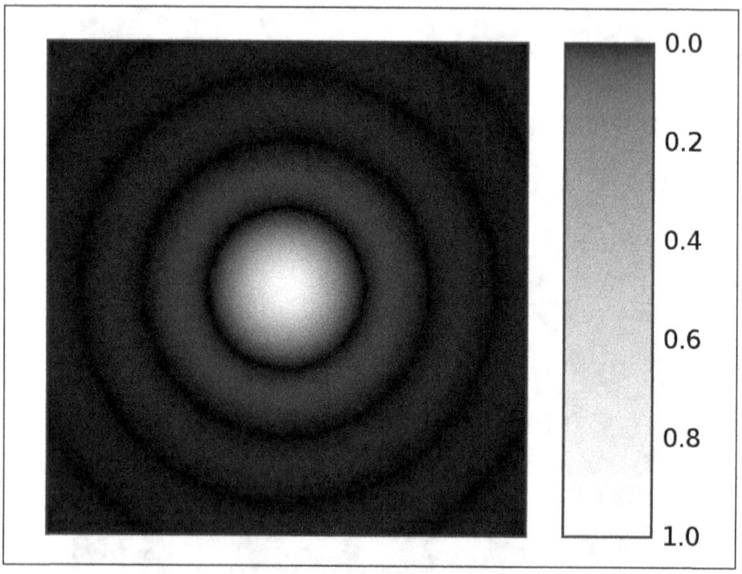

A computer-generated image of an Airy disk.

Computer generated light diffraction pattern from a circular aperture of diameter 0.5 micrometre at a wavelength of 0.6 micrometre (red-light) at distances of 0.1 cm – 1 cm in steps of 0.1 cm. One can see the image moving from the Fresnel region into the Fraunhofer region where the Airy pattern is seen.

Circular Aperture

The far-field diffraction of a plane wave incident on a circular aperture is often referred to as the Airy Disk. The variation in intensity with angle is given by:

$$I(\theta) = I_0 \left(\frac{2J_1(ka \sin\theta)}{ka \sin\theta} \right)^2,$$

where a is the radius of the circular aperture, k is equal to $2\pi/\lambda$ and J_1 is a Bessel function. The smaller the aperture, the larger the spot size at a given distance, and the greater the divergence of the diffracted beams.

General Aperture

The wave that emerges from a point source has amplitude ψ at location r that is given by the solution of the frequency domain wave equation for a point source (the Helmholtz equation),

$$\nabla^2\psi + k^2\psi = \delta(\mathrm{r})$$

where $\delta(\mathrm{r})$ is the 3-dimensional delta function. The delta function has only radial dependence, so the Laplace operator (a.k.a. scalar Laplacian) in the spherical coordinate system simplifies to:

$$\nabla^2\psi = \frac{1}{r}\frac{\partial^2}{\partial r^2}(r\psi)$$

By direct substitution, the solution to this equation can be readily shown to be the scalar Green's function, which in the spherical coordinate system (and using the physics time convention $e^{-i\omega t}$) is:

$$\psi(r) = \frac{e^{ikr}}{4\pi r}$$

This solution assumes that the delta function source is located at the origin. If the source is located at an arbitrary source point, denoted by the vector r' and the field point is located at the point r, then we may represent the scalar Green's function (for arbitrary source location) as:

$$\psi(\mathrm{r}\,|\,\mathrm{r}') = \frac{e^{ik|\mathrm{r}-\mathrm{r}'|}}{4\pi\,|\mathrm{r}-\mathrm{r}'|}$$

Therefore, if an electric field, $E_{inc}(x,y)$ is incident on the aperture, the field produced by this aperture distribution is given by the surface integral:

$$\Psi(r) \propto \iint_{\text{aperture}} E_{\text{inc}}(x', y') \frac{e^{ik|\mathrm{r}-\mathrm{r}'|}}{4\pi\,|\mathrm{r}-\mathrm{r}'|} dx' dy',$$

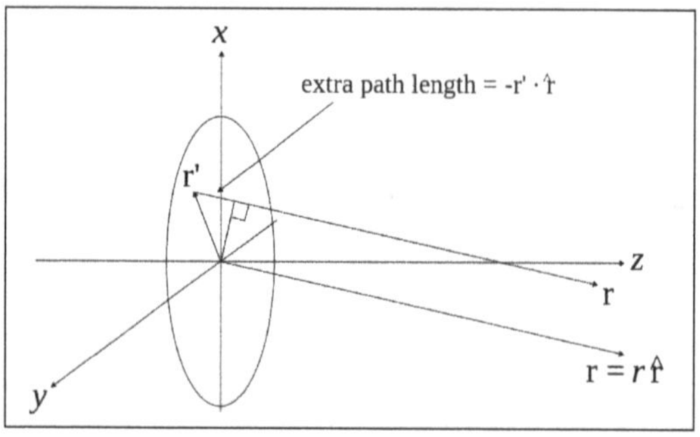

On the calculation of Fraunhofer region fields.

where the source point in the aperture is given by the vector,

$$\mathrm{r}' = x'\,\hat{\mathrm{x}} + y'\,\hat{\mathrm{y}}$$

In the far field, wherein the parallel rays approximation can be employed, the Green's function,

$$\psi(\mathrm{r}\,|\,\mathrm{r}') = \frac{e^{ik|\mathrm{r}-\mathrm{r}'|}}{4\pi\,|\mathrm{r}-\mathrm{r}'|}$$

simplifies to,

$$\psi(\mathrm{r}\,|\,\mathrm{r}') = \frac{e^{ikr}}{4\pi r} e^{-ik(\mathrm{r}'\cdot\hat{\mathrm{r}})}$$

as can be seen in the figure to the right.

The expression for the far-zone (Fraunhofer region) field becomes,

$$\Psi(r) \propto \frac{e^{ikr}}{4\pi r} \iint_{\text{aperture}} E_{\text{inc}}(x', y') e^{-ik(\mathrm{r}'\cdot\hat{\mathrm{r}})}\, dx' dy',$$

Now, since:

$$\mathrm{r}' = x'\,\hat{\mathrm{x}} + y'\,\hat{\mathrm{y}}$$

and,

$$\hat{\mathrm{r}} = \sin\theta\cos\phi\,\hat{\mathrm{x}} + \sin\theta\,\sin\phi\,\hat{\mathrm{y}} + \cos\theta\,\hat{\mathrm{z}}$$

the expression for the Fraunhofer region field from a planar aperture now becomes,

$$\Psi(r) \propto \frac{e^{ikr}}{4\pi r} \iint_{\text{aperture}} E_{\text{inc}}(x', y') e^{-ik\sin\theta(\cos\phi x' + \sin\phi y')} dx' dy'$$

Letting,

$$k_x = k \sin\theta \cos\phi$$

and,

$$k_y = k \sin\theta \sin\phi$$

the Fraunhofer region field of the planar aperture assumes the form of a Fourier transform,

$$\Psi(r) \propto \frac{e^{ikr}}{4\pi r} \iint_{\text{aperture}} E_{\text{inc}}(x', y') e^{-i(k_x x' + k_y y')} dx' dy',$$

In the far-field / Fraunhofer region, this becomes the spatial Fourier transform of the aperture distribution. Huygens' principle when applied to an aperture simply says that the far-field diffraction pattern is the spatial Fourier transform of the aperture shape, and this is a direct by-product of using the parallel-rays approximation, which is identical to doing a plane wave decomposition of the aperture plane fields.

Propagation of a Laser Beam

The way in which the beam profile of a laser beam changes as it propagates is determined by diffraction. When the entire emitted beam has a planar, spatially coherent wave front, it approximates Gaussian beam profile and has the lowest divergence for a given diameter. The smaller the output beam, the quicker it diverges. It is possible to reduce the divergence of a laser beam by first expanding it with one convex lens, and then collimating it with a second convex lens whose focal point is coincident with that of the first lens. The resulting beam has a larger diameter, and hence a lower divergence. Divergence of a laser beam may be reduced below the diffraction of a Gaussian beam or even reversed to convergence if the refractive index of the propagation media increases with the light intensity. This may result in a self-focusing effect.

When the wave front of the emitted beam has perturbations, only the transverse coherence length (where the wave front perturbation is less than 1/4 of the wavelength) should be considered as a Gaussian beam diameter when determining the divergence of the laser beam. If the transverse coherence length in the vertical direction is higher than in horizontal, the laser beam divergence will be lower in the vertical direction than in the horizontal.

Diffraction-limited Imaging

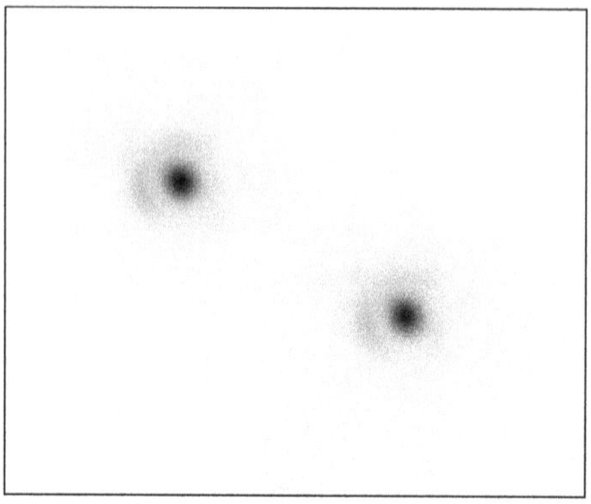

The Airy disk around each of the stars from the 2.56 m telescope aperture can be seen in this lucky image of the binary star zeta Boötis.

The ability of an imaging system to resolve detail is ultimately limited by diffraction. This is because a plane wave incident on a circular lens or mirror is diffracted as described above. The light is not focused to a point but forms an Airy disk having a central spot in the focal plane with radius to first null of,

$$d = 1.22\lambda N,$$

where λ is the wavelength of the light and N is the f-number (focal length divided by diameter) of the imaging optics. In object space, the corresponding angular resolution is,

$$\sin\theta = 1.22\frac{\lambda}{D},$$

where D is the diameter of the entrance pupil of the imaging lens (e.g., of a telescope's main mirror).

Two point sources will each produce an Airy pattern – see the photo of a binary star. As the point sources move closer together, the patterns will start to overlap, and ultimately they will merge to form a single pattern, in which case the two point sources cannot be resolved in the image. The Rayleigh criterion specifies that two point sources can be considered to be resolvable if the separation of the two images is at least the radius of the Airy disk, i.e. if the first minimum of one coincides with the maximum of the other.

Thus, the larger the aperture of the lens, and the smaller the wavelength, the finer the resolution of an imaging system. This is why telescopes have very large lenses or mirrors, and why optical microscopes are limited in the detail which they can see.

Speckle Patterns

The speckle pattern which is seen when using a laser pointer is another diffraction phenomenon. It is a result of the superposition of many waves with different phases, which are produced when a laser beam illuminates a rough surface. They add together to give a resultant wave whose amplitude, and therefore intensity, varies randomly.

Babinet's Principle

Babinet's Principle is a useful theorem stating that the diffraction pattern from an opaque body is identical to that from a hole of the same size and shape, but with differing intensities. This means that the interference conditions of a single obstruction would be the same as that of a single slit.

Nonimaging Optics

Nonimaging optics (also called anidolic optics) is the branch of optics concerned with the optimal transfer of light radiation between a source and a target. Unlike traditional imaging optics, the techniques involved do not attempt to form an image of the source; instead an optimized optical system for optimal radiative transfer from a source to a target is desired.

Applications

The two design problems that nonimaging optics solves better than imaging optics are:

- Solar energy concentration: Maximizing the amount of energy applied to a receiver, typically a solar cell or a thermal receiver.
- Illumination: Controlling the distribution of light, typically so it is "evenly" spread over some areas and completely blocked from other areas.

Typical variables to be optimized at the target include the total radiant flux, the angular distribution of optical radiation, and the spatial distribution of optical radiation. These variables on the target side of the optical system often must be optimized while simultaneously considering the collection efficiency of the optical system at the source.

Solar Energy Concentration

For a given concentration, nonimaging optics provide the widest possible acceptance angles and, therefore, are the most appropriate for use in solar concentration as, for example, in concentrated photovoltaics. When compared to "traditional" imaging optics (such as parabolic reflectors or fresnel lenses), the main advantages of nonimaging optics for concentrating solar energy are:

- Wider acceptance angles resulting in higher tolerances (and therefore higher efficiencies) for:
    - Less precise tracking.

- Imperfectly manufactured optics.
- Imperfectly assembled components.
- Movements of the system due to wind.
- Finite stiffness of the supporting structure.
- Deformation due to aging.
- Capture of circumsolar radiation.
- Other imperfections in the system.

- Higher solar concentrations:
    - Smaller solar cells (in concentrated photovoltaics).
    - Higher temperatures (in concentrated solar thermal).
    - Lower thermal losses (in concentrated solar thermal).
    - Widen the applications of concentrated solar power, for example to solar lasers.
- Possibility of a uniform illumination of the receiver:
    - Improve reliability and efficiency of the solar cells (in concentrated photovoltaics).
    - Improve heat transfer (in concentrated solar thermal).
- Design flexibility: Different kinds of optics with different geometries can be tailored for different applications.

Also, for low concentrations, the very wide acceptance angles of nonimaging optics can avoid solar tracking altogether or limit it to a few positions a year.

The main disadvantage of nonimaging optics when compared to parabolic reflectors or fresnel lenses is that, for high concentrations, they typically have one more optical surface, slightly decreasing efficiency. That, however, is only noticeable when the optics are aiming perfectly towards the sun, which is typically not the case because of imperfections in practical systems.

Illumination Optics

Examples of nonimaging optical devices include optical light guides, nonimaging reflectors, nonimaging lenses or a combination of these devices. Common applications of nonimaging optics include many areas of illumination engineering (lighting). Examples of modern implementations of nonimaging optical designs include automotive headlamps, LCD backlights, illuminated instrument panel displays, fiber optic illumination devices, LED lights, projection display systems and luminaires.

When compared to "traditional" design techniques, nonimaging optics has the following advantages for illumination:

- Better handling of extended sources.
- More compact optics.
- Color mixing capabilities.
- Combination of light sources and light distribution to different places.
- Well suited to be used with increasingly popular LED light sources.
- Tolerance to variations in the relative position of light source and optic.

Examples of nonimaging illumination optics using solar energy are anidolic lighting or solar pipes.

Other Applications

Collecting radiation emitted by high-energy particle collisions using the fewest photomultiplier tubes.

Some of the design methods for nonimaging optics are also finding application in imaging devices, for example some with ultra-high numerical aperture.

Theory

Early academic research in nonimaging optical mathematics seeking closed form solutions was first published in textbook form in a 1978 book. A modern textbook illustrating the depth and breadth of research and engineering in this area was published in 2004. A thorough introduction to this field was published in 2008.

Special applications of nonimaging optics such as Fresnel lenses for solar concentration or solar concentration in general have also been published, although this last reference by O'Gallagher describes mostly the work developed some decades ago.

Imaging optics can concentrate sunlight to, at most, the same flux found at the surface of the sun. Nonimaging optics have been demonstrated to concentrate sunlight to 84,000 times the ambient intensity of sunlight, exceeding the flux found at the surface of the sun, and approaching the theoretical (2nd law of thermodynamics) limit of heating objects up to the temperature of the sun's surface.

The simplest way to design nonimaging optics is called "the method of strings", based on the edge ray principle. Other more advanced methods were developed starting in the early 1990s that can better handle extended light sources than the edge-ray method. These were developed primarily to solve the design problems related to solid state automobile headlamps and complex illumination systems. One of these advanced design methods is the Simultaneous Multiple Surface design method (SMS). The 2D SMS design method (U.S. Patent 6,639,733) is described in detail in the aforementioned textbooks. The 3D SMS design method (U.S. Patent 7,460,985) was developed in 2003 by a team of optical scientists at Light Prescriptions Innovators.

Edge Ray Principle

In simple terms, the edge ray principle states that if the light rays coming from the edges of the source are redirected towards the edges of the receiver, this will ensure that all light rays coming from the inner points in the source will end up on the receiver. There is no condition on image formation, the only goal is to transfer the light from the source to the target.

In figure below edge ray principle on the right illustrates this principle. A lens collects light from a source S_1S_2 and redirects it towards a receiver R_1R_2.

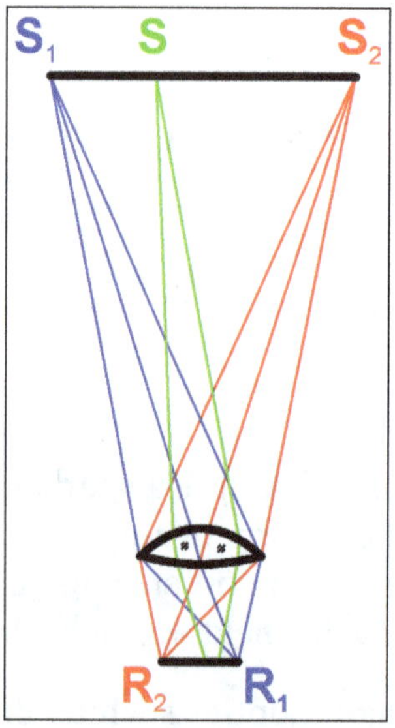

Edge ray principle.

The lens has two optical surfaces and, therefore, it is possible to design it (using the SMS design method) so that the light rays coming from the edge S_1 of the source are redirected towards edge R_1 of the receiver, as indicated by the blue rays. By symmetry, the rays coming from edge S_2 of the source are redirected towards edge R_2 of the receiver, as indicated by the red rays. The rays coming from an inner point S in the source are redirected towards the target, but they are not concentrated onto a point and, therefore, no image is formed.

Actually, if we consider a point P on the top surface of the lens, a ray coming from S_1 through P will be redirected towards R_1. Also a ray coming from S_2 through P will be redirected towards R_2. A ray coming through P from an inner point S in the source will be redirected towards an inner point of the receiver. This lens then guarantees that all light from the source crossing it will be redirected towards the receiver. However, no image of the source is formed on the target. Imposing the condition of image formation on the receiver would imply using more optical surfaces, making the optic more complicated, but would not improve light transfer between source and target (since all light is already transferred). For that reason nonimaging optics are simpler and more efficient than imaging optics in transferring radiation from a source to a target.

Design Methods

Nonimaging optics devices are obtained using different methods. The most important are: the flow-line or Winston-Welford design method, the SMS or Miñano-Benitez design method and the Miñano design method using Poisson brackets. The first (flow-line) is probably the most used, although the second (SMS) has proven very versatile, resulting in a wide variety of optics. The third has remained in the realm of theoretical optics and has not found real world application to date. Often optimization is also used.

Typically optics have refractive and reflective surfaces and light travels through media of different refractive indices as it crosses the optic. In those cases a quantity called optical path length (OPL) may be defined as $S=\sum_i n_i d_i$ where index i indicates different ray sections between successive deflections (refractions or reflections), n_i is the refractive index and d_i the distance in each section i of the ray path.

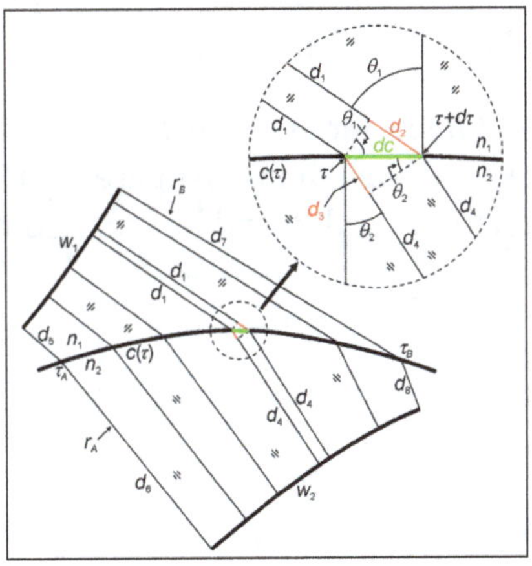

Constant OPL.

The OPL is constant between wavefronts. This can be seen for refraction in the figure "constant OPL" to the right. It shows a separation $c(\tau)$ between two media of refractive indices n_1 and n_2, where $c(\tau)$ is described by a parametric equation with parameter τ. Also shown are a set of rays perpendicular to wavefront w_1 and traveling in the medium of refractive index n_1. These rays refract at $c(\tau)$ into the medium of refractive index n_2 in directions perpendicular to wavefront w_2. Ray r_A crosses c at point $c(\tau_A)$ and, therefore, ray r_A is identified by parameter τ_A on c. Likewise, ray r_B is identified by parameter τ_B on c. Ray r_A has optical path length $S(\tau_A) = n_1 d_5 + n_2 d_6$. Also, ray r_B has optical path length $S(\tau_B) = n_1 d_7 + n_2 d_8$. The difference in optical path length for rays r_A and r_B is given by:

$$S(\tau_B) - S(\tau_A) = \int_A^B dS = \int_{\tau_A}^{\tau_B} \frac{dS}{d\tau} d\tau = \int_{\tau_A}^{\tau_B} \frac{S(\tau + d\tau) - S(\tau)}{(\tau + d\tau) - \tau} d\tau$$

In order to calculate the value of this integral, we evaluate $S(\tau + d\tau) - S(\tau)$, again with the help of the same figure. We have $S(\tau) = n_1 d_1 + n_2(d_3 + d_4)$ and $S(\tau + d\tau) = n_1(d_1 + d_2) + n_2 d_4$. These expressions can be rewritten as $S(\tau) = n_1 d_1 + n_2 dc \sin\theta_2 + n_2 d_4$ and $S(\tau + d\tau) = n_1 d_1 + n_1 dc \sin\theta_1 + n_2 d_4$. From the law of refraction $n_1 \sin\theta_1 = n_2 \sin\theta_2$ and therefore $S(\tau + d\tau) = S(\tau)$, leading to $S(\tau_A) = S(\tau_B)$. Since these may be arbitrary rays crossing c, it may be concluded that the optical path length between w_1 and w_2 is the same for all rays perpendicular to incoming wavefront w_1 and outgoing wavefront w_2.

Similar conclusions may be drawn for the case of reflection, only in this case $n_1 = n_2$. This relationship between rays and wavefronts is valid in general.

Flow-line Design Method

The flow-line (or Winston-Welford) design method typically leads to optics which guide the light confining it between two reflective surfaces. The best known of these devices is the CPC (Compound Parabolic Concentrator).

These types of optics may be obtained, for example, by applying the edge ray of nonimaging optics to the design of mirrored optics, as shown in figure "CEC" on the right. It is composed of two elliptical mirrors e_1 with foci S_1 and R_1 and its symmetrical e_2 with foci S_2 and R_2.

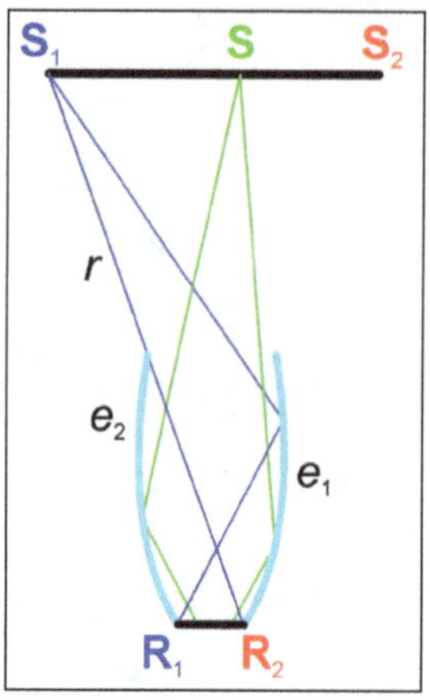

CEC.

Mirror e_1 redirects the rays coming from the edge S_1 of the source towards the edge R_1 of the receiver and, by symmetry, mirror e_2 redirects the rays coming from the edge S_2 of the source towards the edge R_2 of the receiver. This device does not form an image of the source S_1S_2 on the receiver R_1R_2 as indicated by the green rays coming from a point S in the source that end up on the receiver but are not focused onto an image point. Mirror e_2 starts at the edge R_1 of the receiver since leaving a gap between mirror and receiver would allow light to escape between the two. Also, mirror e_2 ends at ray r connecting S_1 and R_2 since cutting it short would prevent it from capturing as much light as possible, but extending it above r would shade light coming from S_1 and its neighboring points of the source. The resulting device is called a CEC (Compound Elliptical Concentrator).

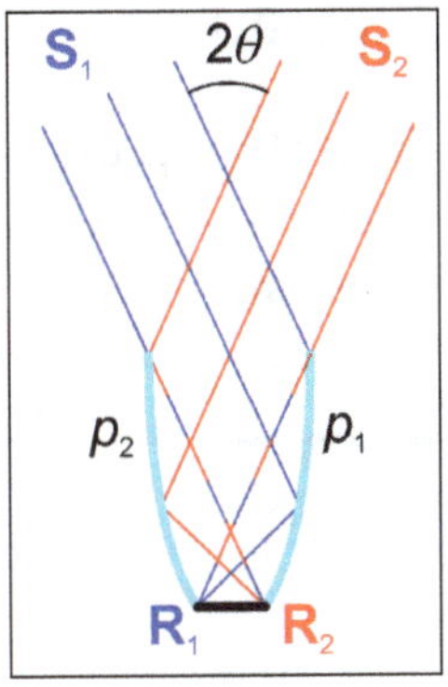

CPC.

A particular case of this design happens when the source S_1S_2 becomes infinitely large and moves to an infinite distance. Then the rays coming from S_1 become parallel rays and the same for those coming from S_2 and the elliptical mirrors e_1 and e_2 converge to parabolic mirrors p_1 and p_2. The resulting device is called a CPC (Compound Parabolic Concentrator), and shown in the "CPC" figure on the left. CPCs are the most common seen nonimaging optics. They are often used to demonstrate the difference between Imaging optics and nonimaging optics.

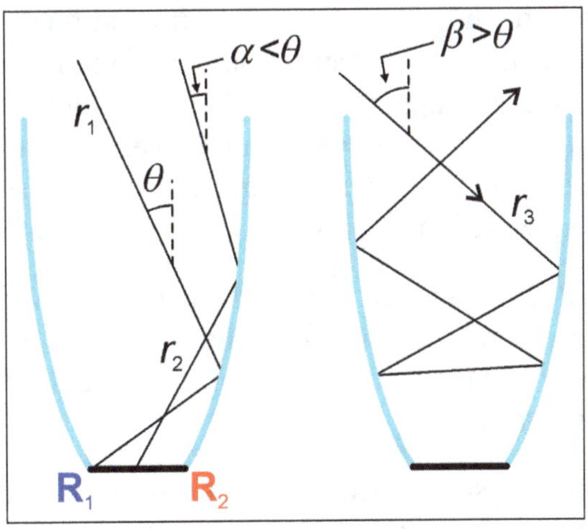

Rays showing the acceptance angle.

When seen from the CPC, the incoming radiation (emitted from the infinite source at an infinite distance) subtends an angle ±θ (total angle 2θ). This is called the acceptance angle of the CPC. The reason for this name can be appreciated in the figure "rays showing the acceptance angle" on the right. An incoming ray r_1 at an angle θ to the vertical (coming from the edge of the infinite source) is redirected by the CPC towards the edge R_1 of the receiver.

Another ray r_2 at an angle $\alpha<\theta$ to the vertical (coming from an inner point of the infinite source) is redirected towards an inner point of the receiver. However, a ray r_3 at an angle $\beta>\theta$ to the vertical (coming from a point outside the infinite source) bounces around inside the CPC until it is rejected by it. Therefore, only the light inside the acceptance angle $\pm\theta$ is captured by the optic; light outside it is rejected.

The ellipses of a CEC can be obtained by the (pins and) string method, as shown in the figure "string method" on the left. A string of constant length is attached to edge point S_1 of the source and edge point R_1 of the receiver.

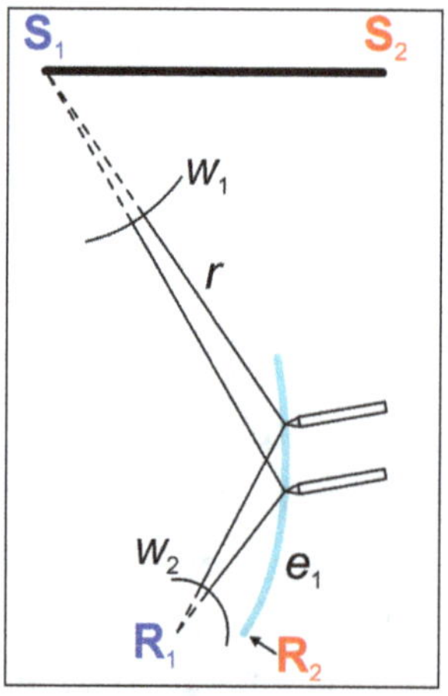

String method

The string is kept stretched while moving a pencil up and down, drawing the elliptical mirror e_1. We can now consider a wavefront w_1 as a circle centered at S_1. This wavefront is perpendicular to all rays coming out of S_1 and the distance from S_1 to w_1 is constant for all its points. The same is valid for wavefront w_2 centered at R_1. The distance from w_1 to w_2 is then constant for all light rays reflected at e_1 and these light rays are perpendicular to both, incoming wavefront w_1 and outgoing wavefront w_2.

Optical path length (OPL) is constant between wavefronts. When applied to nonimaging optics, this result extends the string method to optics with both refractive and reflective surfaces. Figure "DTIRC" (Dielectric Total Internal Reflection Concentrator) on the left shows one such example.

The shape of the top surface *s* is prescribed, for example, as a circle. Then the lateral wall m_1 is calculated by the condition of constant optical path length $S=d_1+n\ d_2+n\ d_3$

where d_1 is the distance between incoming wavefront w_1 and point P on the top surface s, d_2 is the distance between P and Q and d_3 the distance between Q and outgoing wavefront w_2, which is circular and centered at R_1. Lateral wall m_2 is symmetrical to m_1. The acceptance angle of the device is 2θ.

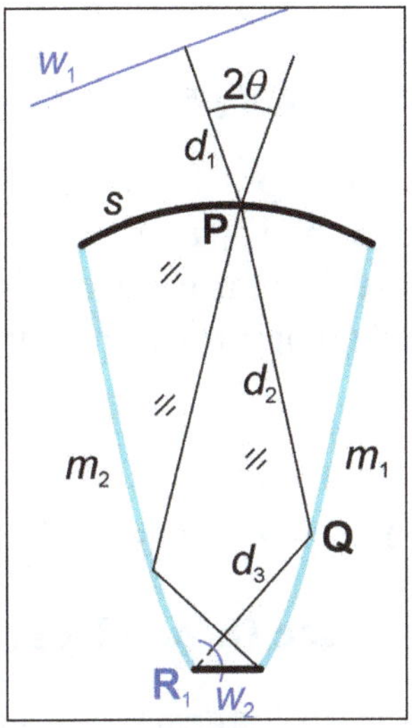

DTIRC.

These optics are called flow-line optics and the reason for that is illustrated in figure "CPC flow-lines" on the right. It shows a CPC with an acceptance angle 2θ, highlighting one of its inner points P.

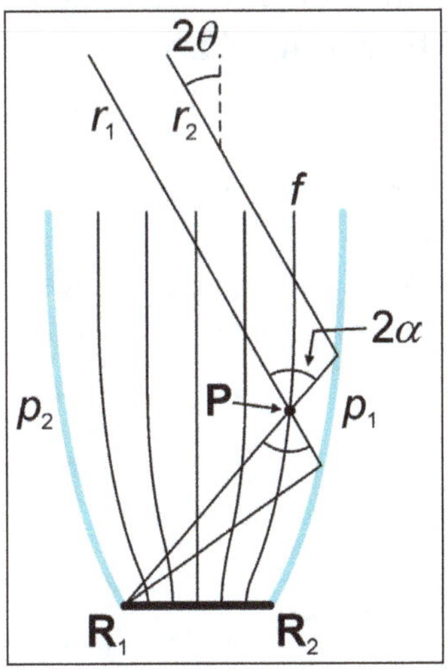

CPC flow-lines.

The light crossing this point is confined to a cone of angular aperture 2α. A line f is also shown whose tangent at point P bisects this cone of light and, therefore, points in the direction of the "light flow" at P. Several other such lines are also shown in the figure. They all bisect the edge rays at each point inside the CPC and, for that reason, their tangent at each point points in the direction of the flow of light. These are called flow-lines and the CPC itself is just a combination of flow line p_1 starting at R_2 and p_2 starting at R_1.

Variations to the Flow-line Design Method

There are some variations to the flow-line design method.

A variation are the multichannel or stepped flow-line optics in which light is split into several "channels" and then recombined again into a single output. Aplanatic (a particular case of SMS) versions of these designs have also been developed. The main application of this method is in the design of ultra-compact optics.

Another variation is the confinement of light by caustics. Instead of light being confined by two reflective surfaces, it is confined by a reflective surface and a caustic of the edge rays. This provides the possibility to add lossless non-optical surfaces to the optics.

Simultaneous Multiple Surface (SMS) Design Method

This topic describes a nonimaging optics design method known in the field as the simultaneous multiple surface (SMS) or the Miñano-Benitez design method. The abbreviation SMS comes from the fact that it enables the simultaneous design of multiple optical surfaces. The original idea came from Miñano. The design method itself was initially developed in 2-D by Miñano and later also by Benítez. The first generalization to 3-D geometry came from Benítez. It was then much further developed by contributions of Miñano and Benítez. Other people have worked initially with Miñano and later with Miñano and Benítez on programming the method.

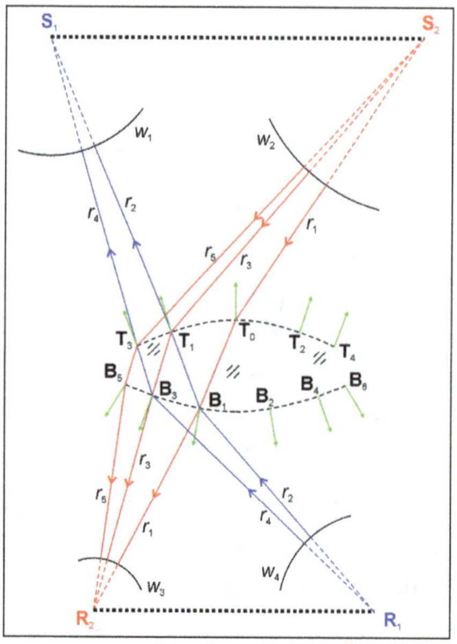

SMS chain.

The design procedure is related to the algorithm used by Schulz in the design of aspheric imaging lenses.

The SMS (or Miñano-Benitez) design method is very versatile and many different types of optics have been designed using it. The 2D version allows the design of two (although more are also possible) aspheric surfaces simultaneously. The 3D version allows the design of optics with freeform surfaces (also called anamorphic) surfaces which may not have any kind of symmetry.

SMS optics are also calculated by applying a constant optical path length between wavefronts. Figure "SMS chain" on the right illustrates how these optics are calculated. In general, the rays perpendicular to incoming wavefront w_1 will be coupled to outgoing wavefront w_4 and the rays perpendicular to incoming wavefront w_2 will be coupled to outgoing wavefront w_3 and these wavefronts may be any shape. However, for the sake of simplicity, this figure shows a particular case or circular wavefronts. This example shows a lens of a given refractive index n designed for a source S_1S_2 and a receiver R_1R_2.

The rays emitted from edge S_1 of the source are focused onto edge R_1 of the receiver and those emitted from edge S_2 of the source are focused onto edge R_2 of the receiver. We first choose a point T_0 and its normal on the top surface of the lens. We can now take a ray r_1 coming from S_2 and refract it at T_0. Choosing now the optical path length S_{22} between S_2 and R_2 we have one condition that allows us to calculate point B_1 on the bottom surface of the lens. The normal at B_1 can also be calculated from the directions of the incoming and outgoing rays at this point and the refractive index of the lens. Now we can repeat the process taking a ray r_2 coming from R_1 and refracting it at B_1. Choosing now the optical path length S_{11} between R_1 and S_1 we have one condition that allows us to calculate point T_1 on the top surface of the lens. The normal at T_1 can also be calculated from the directions of the incoming and outgoing rays at this point and the refractive index of the lens. Now, refracting at T_1 a ray r_3 coming from S_2 we can calculate a new point B_3 and corresponding normal on the bottom surface using the same optical path length S_{22} between S_2 and R_2. Refracting at B_3 a ray r_4 coming from R_1 we can calculate a new point T_3 and corresponding normal on the top surface using the same optical path length S_{11} between R_1 and S_1. The process continues by calculating another point B_5 on the bottom surface using another edge ray r_5, and so on. The sequence of points T_0 B_1 T_1 B_3 T_3 B_5 is called an SMS chain.

Another SMS chain can be constructed towards the right starting at point T_0. A ray from S_1 refracted at T_0 defines a point and normal B_2 on the bottom surface, by using constant optical path length S_{11} between S_1 and R_1. Now a ray from R_2 refracted at B_2 defines a new point and normal T_2 on the top surface, by using constant optical path length S_{22} between S_2 and R_2. The process continues as more points are added to the SMS chain. In this example shown in the figure, the optic has a left-right symmetry and, therefore, points B_2 T_2 B_4 T_4 B_6 can also be obtained by symmetry about the vertical axis of the lens.

Now we have a sequence of spaced points on the plane. Figure "SMS skinning" on the left illustrates the process used to fill the gaps between points, completely defining both optical surfaces.

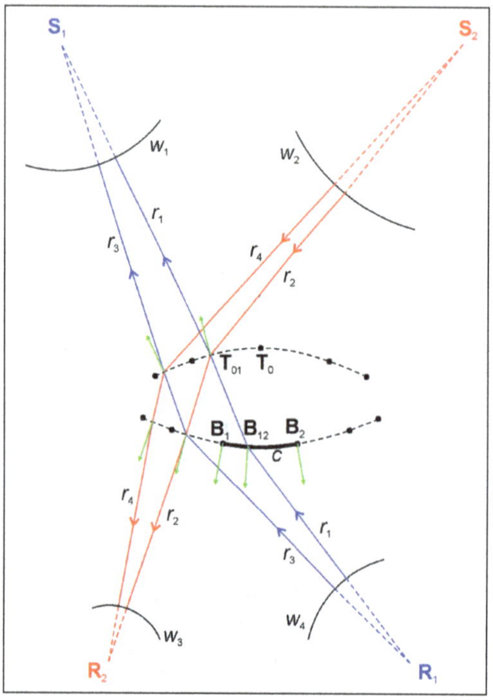

SMS skinning.

We pick two points, say B_1 and B_2, with their corresponding normals and interpolate a curve *c* between them. Now we pick a point B_{12} and its normal on *c*. A ray r_1 coming from R_1 and refracted at B_{12} defines a new point T_{01} and its normal between T_0 and T_1 on the top surface, by applying the same constant optical path length S_{11} between S_1 and R_1. Now a ray r_2 coming from S_2 and refracted at T_{01} defines a new point and normal on the bottom surface, by applying the same constant optical path length S_{22} between S_2 and R_2. The process continues with rays r_3 and r_4 building a new SMS chain filling the gaps between points. Picking other points and corresponding normals on curve *c* gives us more points in between the other SMS points calculated originally.

In general, the two SMS optical surfaces do not need to be refractive. Refractive surfaces are noted R (from Refraction) while reflective surfaces are noted X (from the Spanish word refleXión). Total Internal Reflection (TIR) is noted I. Therefore, a lens with two refractive surfaces is an RR optic, while another configuration with a reflective and a refractive surface is an XR optic. Configurations with more optical surfaces are also possible and, for example, if light is first refracted (R), then reflected (X) then reflected again by TIR (I), the optic is called an RXI.

The SMS 3D is similar to the SMS 2D, only now all calculations are done in 3D space. Figure "SMS 3D chain" on the right illustrates the algorithm of an SMS 3D calculation.

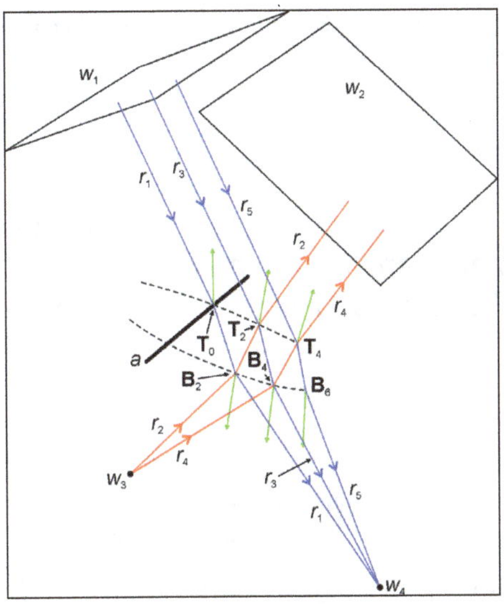

SMS 3D chain.

The first step is to choose the incoming wavefronts w_1 and w_2 and outgoing wavefronts w_3 and w_4 and the optical path length S_{14} between w_1 and w_4 and the optical path length S_{23} between w_2 and w_3. In this example the optic is a lens (an RR optic) with two refractive surfaces, so its refractive index must also be specified. One difference between the SMS 2D and the SMS 3D is on how to choose initial point T_0, which is now on a chosen 3D curve a. The normal chosen for point T_0 must be perpendicular to curve a. The process now evolves similarly to the SMS 2D. A ray r_1 coming from w_1 is refracted at T_0 and, with the optical path length S_{14}, a new point B_2 and its normal is obtained on the bottom surface. Now ray r_2 coming from w_3 is refracted at B_2 and, with the optical path length S_{23}, a new point T_2 and its normal is obtained on the top surface. With ray r_3 a new point B_2 and its normal are obtained, with ray r_4 a new point T_4 and its normal are obtained, and so on. This process is performed in 3D space and the result is a 3D SMS chain. As with the SMS 2D, a set of points and normals to the left of T_0 can also be obtained using the same method. Now, choosing another point T_0 on curve a the process can be repeated and more points obtained on the top and bottom surfaces of the lens.

The power of the SMS method lies in the fact that the incoming and outgoing wavefronts can themselves be free-form, giving the method great flexibility. Also, by designing optics with reflective surfaces or combinations of reflective and refractive surfaces, different configurations are possible.

Miñano Design Method using Poisson Brackets

This design method was developed by Miñano and is based on Hamiltonian optics, the Hamiltonian formulation of geometrical optics which shares much of the mathematical formulation with Hamiltonian mechanics. It allows the design of optics with variable refractive index, and therefore solves some nonimaging problems that are not solvable using other methods. However, manufacturing of variable refractive index optics is still not possible and this method, although potentially powerful, did not yet find a practical application.

Conservation of Etendue

Conservation of etendue is a central concept in nonimaging optics. In concentration optics, it relates the acceptance angle with the maximum concentration possible. Conservation of etendue may be seen as constant a volume moving in phase space.

Köhler Integration

In some applications it is important to achieve a given irradiance (or illuminance) pattern on a target, while allowing for movements or inhomogeneities of the source. Figure "Köhler integrator" on the right illustrates this for the particular case of solar concentration. Here the light source is the sun moving in the sky. On the left this figure shows a lens L_1 L_2 capturing sunlight incident at an angle α to the optical axis and concentrating it onto a receiver L_3 L_4. As seen, this light is concentrated onto a hotspot on the receiver. This may be a problem in some applications. One way around this is to add a new lens extending from L_3 to L_4 that captures the light from L_1 L_2 and redirects it onto a receiver R_1 R_2, as shown in the middle of the figure.

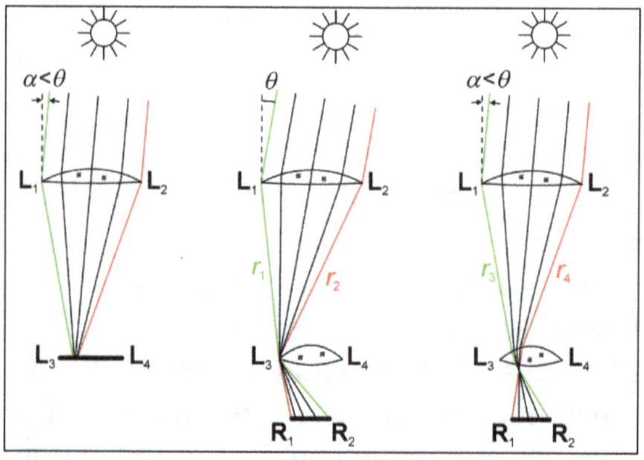

Köhler integrator.

The situation in the middle of the figure shows a nonimaging lens L_1 L_2 is designed in such a way that sunlight (here considered as a set of parallel rays) incident at an angle θ to the optical axis will be concentrated to point L_3. On the other hand, nonimaging lens L_3 L_4 is designed in such a way that light rays coming from L_1 are focused on R_2 and light rays coming from L_2 are focused on R_1. Therefore, ray r_1 incident on the first lens at an angle θ will be redirected towards L_3. When it hits the second lens, it is coming from point L_1 and it is redirected by the second lens to R_2. On the other hand, ray r_2 also incident on the first lens at an angle θ will also be redirected towards L_3. However, when it hits the second lens, it is coming from point L_2 and it is redirected by the second lens to R_1. Intermediate rays incident on the first lens at an angle θ will be redirected to points between R_1 and R_2, fully illuminating the receiver.

Something similar happens in the situation shown in the same figure, on the right. Ray r_3 incident on the first lens at an angle $\alpha<\theta$ will be redirected towards a point between L_3 and L_4. When it hits the second lens, it is coming from point L_1 and it is redirected by the second lens to R_2. Also, Ray r_4 incident on the first lens at an angle $\alpha<\theta$ will be redirected towards a point between L_3 and L_4. When it hits the second lens, it is coming from point L_2 and it is redirected by the second lens to R_1. Intermediate rays incident on the first lens at an angle $\alpha<\theta$ will be redirected to points between R_1 and R_2, also fully illuminating the receiver.

This combination of optical elements is called Köhler illumination. Although the example given here was for solar energy concentration, the same principles apply for illumination in general. In practice, Köhler optics are typically not designed as a combination of nonimaging optics, but they are simplified versions with a lower number of active optical surfaces. This decreases the effectiveness of the method, but allows for simpler optics. Also, Köhler optics are often divided into several sectors, each one of them channeling light separately and then combining all the light on the target.

An example of one of these optics used for solar concentration is the Fresnel-R Köhler.

Compound Parabolic Concentrator

In the drawing opposite there are two parabolic mirrors CC' (red) and DD' (blue). Both parabolas are cut at B and A respectively. A is the focal point of parabola CC' and B is the focal point of the parabola DD' The area DC is the entrance aperture and the flat absorber is AB. This CPC has an acceptance angle of θ.

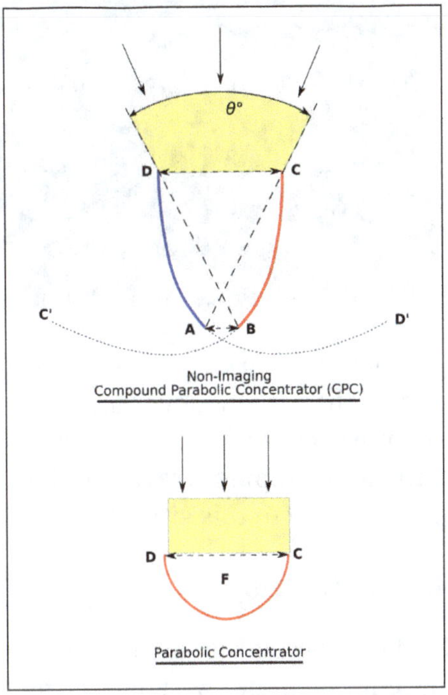

Comparison between non-imaging compound parabolic concentrator and parabolic concentrator.

The parabolic concentrator has an entrance aperture of DC and a focal point F.

The parabolic concentrator only accepts rays of light that are perpendicular to the entrance aperture DC. The tracking of this type of concentrator must be more exact and requires expensive equipment.

The compound parabolic concentrator accepts a greater amount of light and needs less accurate tracking.

For a 3-dimensional "nonimaging compound parabolic concentrator", the maximum concentration possible in air or in vacuum (equal to the ratio of input and output aperture areas), is:

$$C = \frac{1}{\sin^2 \theta},$$

where θ is the half-angle of the acceptance angle (of the larger aperture).

Doppler Effect in Light

Doppler Effect in Light: Red and Blue Shift

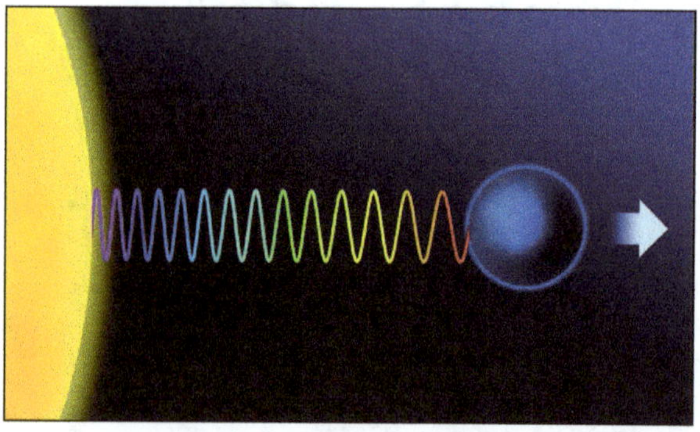

Light waves from a moving source experience the Doppler effect to result in either a red shift or blue shift in the light's frequency. This is in a fashion similar (though not identical) to other sorts of waves, such as sound waves. The major difference is that light waves do not require a medium for travel, so the classical application of the Doppler effect doesn't apply precisely to this situation.

Relativistic Doppler Effect for Light

Consider two objects: the light source and the "listener" (or observer). Since light waves traveling in empty space have no medium, we analyze the Doppler effect for light in terms of the motion of the source relative to the listener.

We set up our coordinate system so that the positive direction is from the listener toward the source. So if the source is moving away from the listener, its velocity v is positive, but if it is moving toward the listener, then the v is negative. The listener, in this case, is always considered to be at rest (so v is really the total relative velocity between them). The speed of light c is always considered positive.

The listener receives a frequency f_L which would be different from the frequency transmitted by the source f_S. This is calculated with relativistic mechanics, by applying necessary the length contraction, and obtains the relationship:

$$f_L = sqrt\left[(c-v)/(c+v)\right] * f_S$$

Red Shift and Blue Shift

A light source moving away from the listener (v is positive) would provide an f_L that is less than f_S. In the visible light spectrum, this causes a shift toward the red end of the light spectrum, so it is called a redshift. When the light source is moving toward the listener (v is negative), then f_L is greater than f_S. In the visible light spectrum, this causes a shift toward the high-frequency end of the light spectrum. For some reason, violet got the short end of the stick and such frequency shift is actually called a blue shift. Obviously, in the area of the electromagnetic spectrum outside of the visible light spectrum, these shifts might not actually be toward red and blue. If you're in the infrared, for example, you're ironically shifting away from red when you experience a "redshift".

Applications

Police use this property in the radar boxes they use to track speed. Radio waves are transmitted out, collide with a vehicle, and bounce back. The speed of the vehicle (which acts as the source of the reflected wave) determines the change in frequency, which can be detected with the box. (Similar applications can be used to measure wind velocities in the atmosphere, which is the "Doppler radar" of which meteorologists are so fond).

This Doppler shift is also used to track satellites. By observing how the frequency changes, you can determine the velocity relative to your location, which allows ground-based tracking to analyze the movement of objects in space.

References

- Coherent-and-incoherent-addition-of-waves, wave-optics, physics, guides: toppr.com, Retrieved 1 August, 2019
- University Physics – With Modern Physics (12th Edition), H. D. Young, R. A. Freedman (Original edition), Addison-Wesley (Pearson International), 1st Edition: 1949, 12th Edition: 2008, ISBN 0-321-50130-6, ISBN 978-0-321-50130-1
- Wave-theory-of-light, physics: byjus.com, Retrieved 2 January, 2019

- Arumugam, Nadia. "Food Explainer: Why Is Some Deli Meat Iridescent?". Slate. The Slate Group. Archived from the original on 10 September 2013. Retrieved 9 September 2013
- Interference-light-waves-youngs-experiment, wave-optics, physics, study-material: emedicalprep.com, Retrieved 3 February, 2019
- Halliday, David; Resnick, Robert; Walker, Jerl (2005), Fundamental of Physics (7th ed.), USA: John Wiley and Sons, Inc., ISBN 978-0-471-23231-5
- Polarization, light: physicsclassroom.com, Retrieved 4 March, 2019
- Diffraction-of-light: physicsabout.com, Retrieved 5 April, 2019

4

Light as an Electromagnetic Wave

The electromagnetic wave nature of light includes the Maxwell's equations, Malus' law, the electromagnetic spectrum, polarization of light, etc. This chapter closely examines these key concepts associated with the electromagnetic wave nature of light to provide an extensive understanding of the subject.

In spite of theoretical and experimental advances in the first half of the 19th century that established the wave properties of light, the nature of light was not yet revealed—the identity of the wave oscillations remained a mystery. This situation dramatically changed in the 1860s when the Scottish physicist James Clerk Maxwell, in a watershed theoretical treatment, unified the fields of electricity, magnetism, and optics. In his formulation of electromagnetism, Maxwell described light as a propagating wave of electric and magnetic fields. More generally, he predicted the existence of electromagnetic radiation: coupled electric and magnetic fields traveling as waves at a speed equal to the known speed of light. In 1888 German physicist Heinrich Hertz succeeded in demonstrating the existence of long-wavelength electromagnetic waves and showed that their properties are consistent with those of the shorter-wavelength visible light.

Electric and Magnetic Fields

The subjects of electricity and magnetism were well developed by the time Maxwell began his synthesizing work. English physician William Gilbert initiated the careful study of magnetic phenomena in the late 16th century. In the late 1700s an understanding of electric phenomena was pioneered by Benjamin Franklin, Charles-Augustin de Coulomb, and others. Siméon-Denis Poisson, Pierre-Simon Laplace, and Carl Friedrich Gauss developed powerful mathematical descriptions of electrostatics and magnetostatics that stand to the present time. The first connection between electric and magnetic effects was discovered by Danish physicist Hans Christian Ørsted in 1820 when he found that electric currents produce magnetic forces. Soon after, French physicist André-Marie Ampère developed a mathematical formulation (Ampère's law) relating currents to magnetic effects. In 1831 the great English experimentalist Michael Faraday discovered electromagnetic induction, in which a moving magnet (more generally, a changing magnetic flux) induces an electric current in a conducting circuit.

Faraday's conception of electric and magnetic effects laid the groundwork for Maxwell's equations. Faraday visualized electric charges as producing fields that extend through space and transmit electric and magnetic forces to other distant charges. The notion of electric and magnetic fields is central to the theory of electromagnetism, and so it requires some explanation. A field is used to represent any physical quantity whose value changes from one point in space to another. For example, the temperature of Earth's atmosphere has a definite value at every point above the surface of Earth; to specify the atmospheric temperature completely thus requires specifying a distribution of numbers—one for each spatial point. The temperature "field" is simply a mathematical accounting of those numbers; it may be expressed as a function of the spatial coordinates. The values of the temperature field can also vary with time; therefore, the field is more generally expressed as a function of spatial coordinates and time: T(x, y, z, t), where T is the temperature field, x, y, and z are the spatial coordinates, and t is the time.

Temperature is an example of a scalar field; its complete specification requires only one number for each spatial point. Vector fields, on the other hand, describe physical quantities that have a direction and magnitude at each point in space. A familiar example is the velocity field of a fluid. Electric and magnetic fields are also vector fields; the electric field is written as $E(x, y, z, t)$ and the magnetic field as $B(x, y, z, t)$.

The Electromagnetic Spectrum

Electromagnetic waves can be classified and arranged according to their various wavelengths/frequencies; this classification is known as the electromagnetic spectrum. The following table shows us this spectrum, which consists of all the types of electromagnetic radiation that exist in our universe.

As we can see, the visible spectrum—that is, light that we can see with our eyes—makes up only a small fraction of the different types of radiation that exist. To the right of the visible spectrum, we find the types of energy that are lower in frequency (and thus longer in wavelength) than visible light. These types of energy include infrared (IR) rays (heat waves given off by thermal bodies), microwaves, and radio waves. These types of radiation surround us constantly, and are not harmful, because their frequencies are so low. As we will see in the section, "the photon," lower frequency waves are lower in energy, and thus are not dangerous to our health.

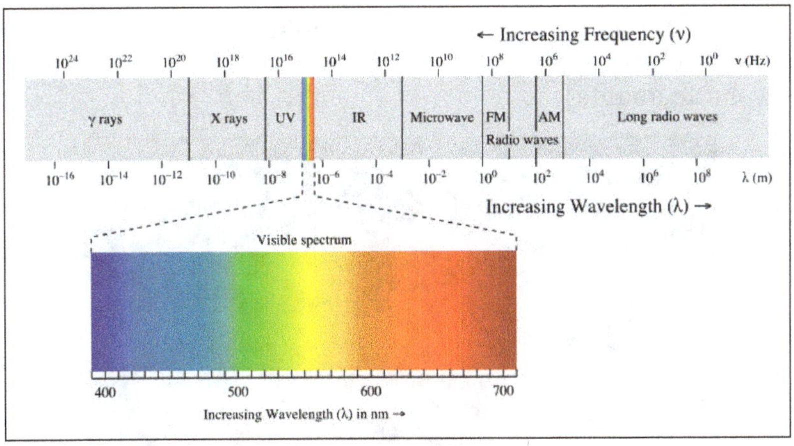

The electromagnetic spectrum.

To the left of the visible spectrum, we have ultraviolet (UV) rays, X-rays, and gamma rays. These types of radiation are harmful to living organisms, due to their extremely high frequencies (and thus, high energies). It is for this reason that we wear suntan lotion at the beach (to block the UV rays from the sun) and why an X-ray technician will place a lead shield over us, in order to prevent the X-rays from penetrating anything other than the area of our body being imaged. Gamma rays, being the highest in frequency and energy, are the most damaging. Luckily though, our atmosphere absorbs gamma rays from outer space, thereby protecting us from harm.

Quantization of Energy and the Dual Nature of Light

We have already described how light travels through space as a wave. This has been well-known for quite some time; in fact, the Dutch physicist Christiaan Huygens first described the wave nature of light as far back as the late seventeenth century. For about 200200200 years after Huygens, physicists assumed that light waves and matter were quite distinct from one another. According to classical physics, matter was composed of particles that had mass, and whose position in space could be known; light waves, on the other hand, were considered to have zero mass, and their position in space could not be determined. Because they were considered to be in different categories, scientists did not have a good understanding of how light and matter interacted. This all changed in 190019001900, however, when the physicist Max Planck began studying blackbodies – bodies heated until they began to glow.

Planck found that the electromagnetic radiation emitted by blackbodies could not be explained by classical physics, which postulated that matter could absorb or emit any quantity of electromagnetic radiation. Planck observed that matter actually absorbed or emitted energy only in whole-number multiples of the value $h\nu$, where h is Planck's constant, 6.626×10^{-34} $J \cdot s$, and ν is the frequency of the light absorbed or emitted. This was a shocking discovery, because it challenged the idea that energy was continuous, and could be transferred in any amount. The reality, which Planck discovered, is that energy is not continuous but quantized—meaning that it can only be transferred in individual "packets" (or particles) of the size hv. Each of these energy packets is known as a quantum (plural: quanta).

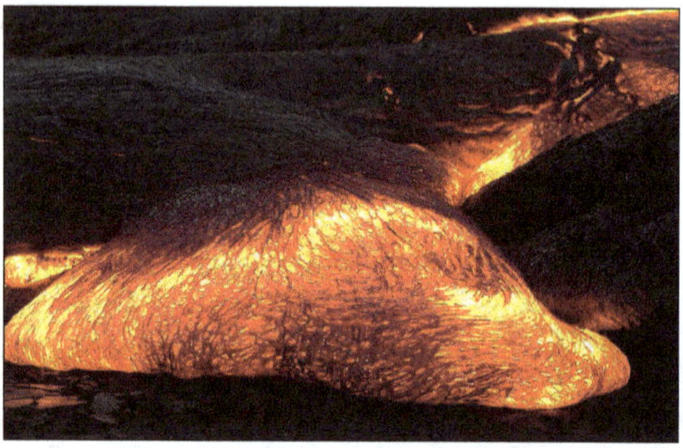

Molten lava emitting blackbody radiation.

While this might sound confusing, we are actually already very familiar with quantized systems. The money we use daily, for example, is quantized. For instance, when you go into a store, you will not see anything on sale for a price of one dollar and two and a half cents $1.025) left parenthesis, dollar sign, 1, point, 025, right parenthesis. This is because the smallest possible monetary unit is the penny—it is impossible to transfer money in any smaller amount than this. Just as we cannot pay the cashier at the store half of a cent, energy cannot be transferred in anything less than a single quantum. We can think of quanta as being like "pennies" of electromagnetic energy—the smallest possible units by which such energy can be transferred.

Planck's discovery that electromagnetic radiation is quantized forever changed the idea that light behaves purely as a wave. In actuality, light seemed to have both wavelike and particle-like properties.

The Photon

Planck's discoveries paved the way for the discovery of the photon. A photon is the elementary particle, or quantum, of light. As we will soon see, photons can be absorbed or emitted by atoms and molecules. When a photon is absorbed, its energy is transferred to that atom or molecule. Because energy is quantized, the photon's entire energy is transferred (remember that we cannot transfer fractions of quanta, which are the smallest possible individual "energy packets"). The reverse of this process is also true. When an atom or molecule loses energy, it emits a photon that carries an energy exactly equal to the loss in energy of the atom or molecule. This change in energy is directly proportional to the frequency of photon emitted or absorbed. This relationship is given by Planck's famous equation:

$$E = hv$$

where E is the energy of the photon absorbed or emitted (given in Joules, J), v is frequency of the photon (given in Hertz, Hz), and h is Planck's constant, $6.626 \times 10^{-34}\ J \cdot s$.

The Maxwell's Equations

Light is an electromagnetic wave. To explain this, we consider Maxwell's equations, the four equations that describe all of electricity and magnetism. The first equation is Ampere's law, only that the second term on the right-hand-side is new. The second equation is Lenz's law. The third equation is Gauss's law and expresses the fact that electric field lines begin and end only at charges. The fourth equation is the magnetic equivalent of Gauss's law, expressing tha fact that magnetic field lines never begin or end.

$$\sum_i \vec{B}_i \cdot \vec{\ell}_i = \mu_0 \left(I + \varepsilon_0 \frac{\Delta E \cdot A}{\Delta t} \right)$$

$$\sum_i \vec{E}_i \cdot \vec{\ell}_i = -\frac{\Delta B \cdot A}{\Delta t}$$

$$\sum_i E_i \cdot A_i = \frac{Q}{\varepsilon_0}$$

$$\sum_i B_i \cdot A_i = 0$$

After a bit of non-trivial differential calculus one can show that both the electric and magnetic field obey wave equations. The speed **c** of an electromagnetic wave is given by: $c^2 = \mathbf{1}/(e_o m_o) = 2.998\ m/s$.

The details of an electromagnetic wave are fairly complicated. If the wave moves in z-direction, the electric field oscillates in the plane perpendicular to the z-axis. The magnetic field also oscillates and is perpendicular to both the magnetic and electric fields. If the electric field is oscillating in the x-direction, we say the wave is polarized along the x-axis. The figure below illustrates the behavior of an electromagnetic wave which is polarized along the x-axis.

The forms of the electric and magnetic field in the wave are:

$$E_x(z,t) = E_{\max} \cos 2\pi(ft - z/\lambda)$$
$$B_y(z,t) = B_{\max} \cos 2\pi(ft - z/\lambda)$$
$$E_{\max} = cB_{\max}$$

where the frequency and wavelength are related by $c = fl$.

Polarization of Light

Light is an electromagnetic wave, and the electric field of this wave oscillates perpendicularly to the direction of propagation. Light is called unpolarized if the direction of this electric field fluctuates randomly in time. Many common light sources such as sunlight, halogen lighting, LED spotlights, and incandescent bulbs produce unpolarized light. If the direction of the electric field of light is well defined, it is called polarized light. The most common source of polarized light is a laser.

Depending on how the electric field is oriented, we classify polarized light into three types of polarizations:

- Linear polarization: The electric field of light is confined to a single plane along the direction of propagation.
- Circular polarization: The electric field of light consists of two linear components that are perpendicular to each other, equal in amplitude, but have a phase difference of $\pi/2$. The resulting electric field rotates in a circle around the direction of propagation and, depending on the rotation direction, is called left- or right-hand circularly polarized light.
- Elliptical polarization: The electric field of light describes an ellipse. This results from the combination of two linear components with differing amplitudes and/or a phase difference that is not $\pi/2$. This is the most general description of polarized light, and circular and linear polarized light can be viewed as special cases of elliptically polarized light.

The two orthogonal linear polarization states that are most important for reflection and transmission are referred to as p- and s-polarization. P-polarized (from the German parallel) light has an electric field polarized parallel to the plane of incidence, while s-polarized (from the German senkrecht) light is perpendicular to this plane.

Malus' Law

At the beginning of the nineteenth century the only known way to generate polarized light was with a calcite crystal. In 1808, using a calcite crystal, Malus discovered that natural incident light became polarized when it wasreflected by a glass surface, and that the light reflected close to an angle of incidence of 57° could be extinguished when viewed through the crystal. He then proposed that natural light consisted of the *s*- and *p*-polarizations, which were perpendicular to each other.

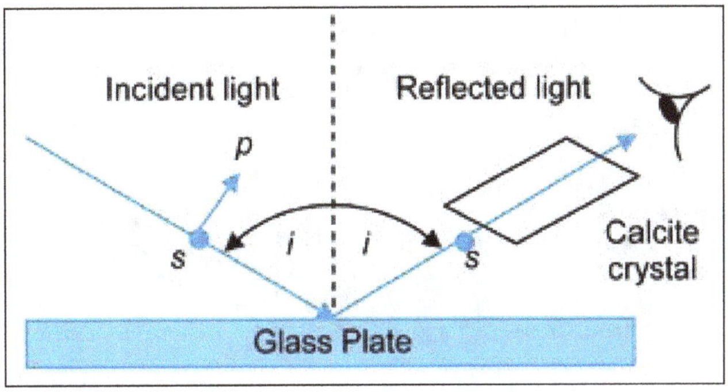

Since the intensity of the reflected light varied from a maximum to a minimum as the crystal was rotated, Malus proposed that the amplitude of the reflected beammust be $A = A_0 \cos\theta$. However, in order to obtain the intensity, Malus squared the amplitude relation so that the intensity equation $I(\theta)$ of the reflected polarized light was:

$$I(\theta) = I_0 \cos^2 \theta$$

where $I_0 = A_0^2$. this equation is known as Malus's Law.

References

- Light-as-electromagnetic-radiation, light, science: britannica.com, Retrieved 6 May, 2019
- Light-and-the-electromagnetic-spectrum, introduction-to-light-waves, light-waves, physics, science: khanacademy.org, Retrieved 7 June, 2019
- Introduction-to-polarization, optics, application-notes, resources: edmundoptics.com, Retrieved 9 August, 2019
- Maxwell, emwaves, lectures, pratts, people: web.pa.msu.edu, Retrieved 8 July, 2019

5
Particle Nature of the Light

The photoelectric effect refers to the emission of electrons when a ray of light strikes a material. Compton effect is the scattering of photons by an electron which results in the decrease of energy of the photons. This chapter delves into the subject related to the particle nature of light for an in-depth understanding.

Light behaves mainly like a wave but it can also be considered to consist of tiny packages of energy called photons. Photons carry a fixed amount of energy but have no mass. The energy of a photon depends on its wavelength: longer wavelength photons have less energy and shorter wavelength photons have more. Red photons, for example, have less energy than blue ones.

Until about 1900, scientists only understood electromagnetic radiation to be made up of waves. Then Max Planck and others were studying the photoelectric effect and they found that certain types of metal and other materials will eject electrons when light shines on them. They expected that the number of electrons ejected from the metal would increase with the intensity or brightness of the light directed towards the metal. What they found instead, was that the wavelength of the light was what affected the number of electrons ejected.

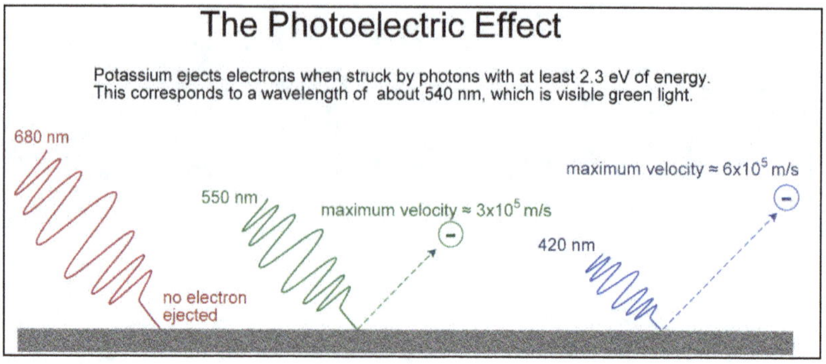

More energetic wavelengths such as blue and ultraviolet caused more electrons to be ejected than red or infrared wavelengths. They also found that increasing the intensity of light increased the number of electrons ejected, but not their speed. Planck realized that the energy of the electromagnetic radiation was proportional to its frequency, but admitted that he didn't understand why this was the case and said it was lucky guesswork.

Einstein was the first to explain what was happening. He theorized that electromagnetic energy comes in packets, or quanta which we now call photons. So light behaves as a wave and as a particle, depending on the circumstances and the effect being observed. This concept is now known as wave-particle duality. Einstein won the 1921 Nobel Prize in Physics "for his services to Theoretical Physics, and especially for his discovery of the law of the photoelectric effect".

The Photoelectric Effect

When light shines on a metal, electrons can be ejected from the surface of the metal in a phenomenon known as the photoelectric effect. This process is also often referred to as photoemission, and the electrons that are ejected from the metal are called photoelectrons. In terms of their behavior and their properties, photoelectrons are no different from other electrons. The prefix, photo-, simply tells us that the electrons have been ejected from a metal surface by incident light.

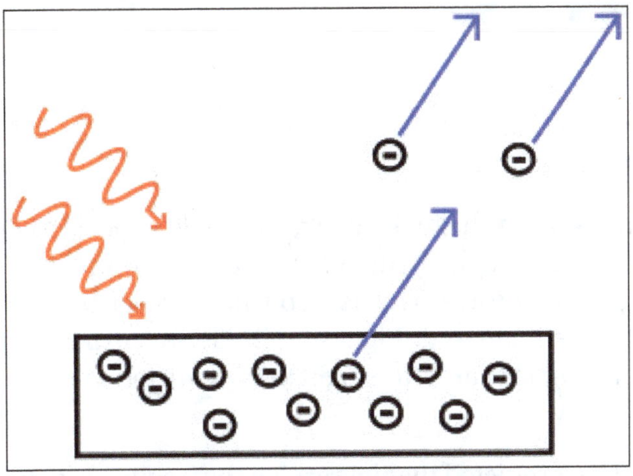

In the photoelectric effect, light waves (red wavy lines) hitting a metal surface cause electrons to be ejected from the metal.

Predictions based on Light as a Wave

To explain the photoelectric effect, 19th-century physicists theorized that the oscillating electric field of the incoming light wave was heating the electrons and causing them to vibrate, eventually freeing them from the metal surface. This hypothesis was based on the assumption that light traveled purely as a wave through space. Scientists also believed that the energy of the light wave was proportional to its brightness, which is related to the wave's amplitude. In order to test their hypotheses, they performed experiments to look at the effect of light amplitude and frequency on the rate of electron ejection, as well as the kinetic energy of the photoelectrons.

Based on the classical description of light as a wave, they made the following predictions:

- The kinetic energy of emitted photoelectrons should increase with the light amplitude.
- The rate of electron emission, which is proportional to the measured electric current, should increase as the light frequency is increased.

To help us understand why they made these predictions, we can compare a light wave to a water wave. Imagine some beach balls sitting on a dock that extends out into the ocean. The dock represents a metal surface, the beach balls represent electrons, and the ocean waves represent light waves.

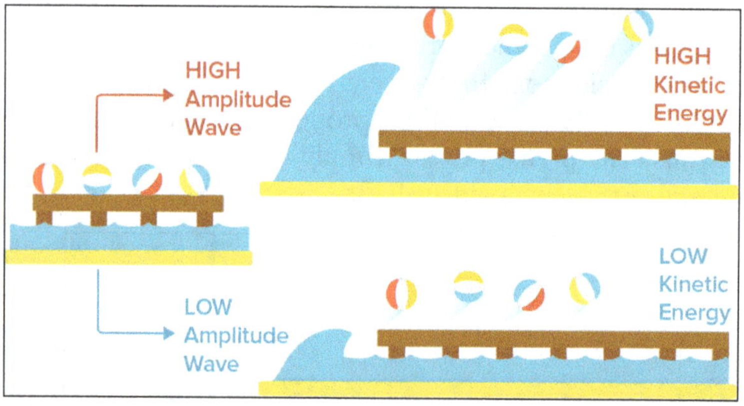

If a single large wave were to shake the dock, we would expect the energy from the big wave would send the beach balls flying off the dock with much more kinetic energy compared to a single, small wave. This is also what physicists believed would happen if the light intensity was increased. Light amplitude was expected to be proportional to the light energy, so higher amplitude light was predicted to result in photoelectrons with more kinetic energy.

Classical physicists also predicted that increasing the frequency of light waves (at a constant amplitude) would increase the rate of electrons being ejected, and thus increase the measured electric current. Using our beach ball analogy, we would expect waves hitting the dock more frequently would result in more beach balls being knocked off the dock compared to the same sized waves hitting the dock less often.

Now that we know what physicists thought would happen, let's look at what they actually observed experimentally!

When intuition Fails: Photons to the Rescue

When experiments were performed to look at the effect of light amplitude and frequency, the following results were observed:

- The kinetic energy of photoelectrons increases with light frequency.
- Electric current remains constant as light frequency increases.
- Electric current increases with light amplitude.
- The kinetic energy of photoelectrons remains constant as light amplitude increases.

These results were completely at odds with the predictions based on the classical description of light as a wave! In order to explain what was happening, it turned out that an entirely new model of light was needed. That model was developed by Albert Einstein, who proposed that light sometimes behaved as particles of electromagnetic energy which we now call photons. The energy of a photon could be calculated using Planck's equation:

$$E_{\text{photon}} = h\nu$$

Where E_{photon} is the energy of a photon in joules (J) h is Planck's constant (6.626×10^{-34} J·s) and ν is the frequency of the light in Hz. According to Planck's equation, the energy of a photon is proportional to the frequency of the light, ν. The amplitude of the light is then proportional to the number of photons with a given frequency.

Light Frequency and the Threshold Frequency ν_0

We can think of the incident light as a stream of photons with an energy determined by the light frequency. When a photon hits the metal surface, the photon's energy is absorbed by an electron in the metal. The graphic below illustrates the relationship between light frequency and the kinetic energy of ejected electrons.

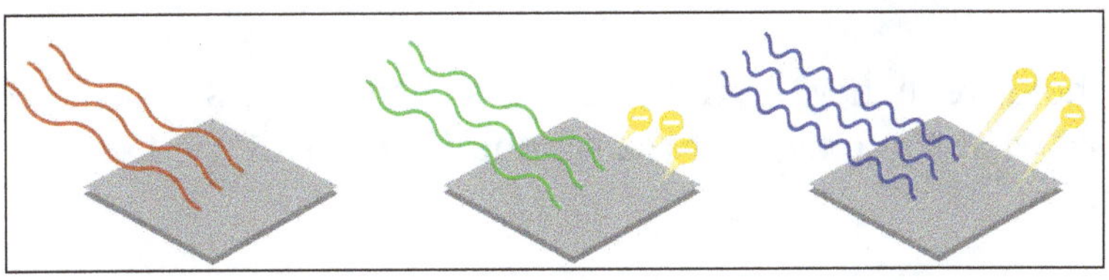

The frequency of red light (left) is less than the threshold frequency of this metal ($\nu_{\text{red}} < \nu_0$), so no electrons are ejected. The green (middle) and blue light (right) have $\nu > \nu_0$, so both cause photoemission. The higher energy blue light ejects electrons with higher kinetic energy compared to the green light.

The scientists observed that if the incident light had a frequency less than a minimum frequency v_0, then no electrons were ejected regardless of the light amplitude. This minimum frequency is also called the threshold frequency, and the value of v_0 depends on the metal. For frequencies greater than v_0, electrons would be ejected from the metal. Furthermore, the kinetic energy of the photoelectrons was proportional to the light frequency. The relationship between photoelectron kinetic energy and light frequency is shown in graph (a) below.

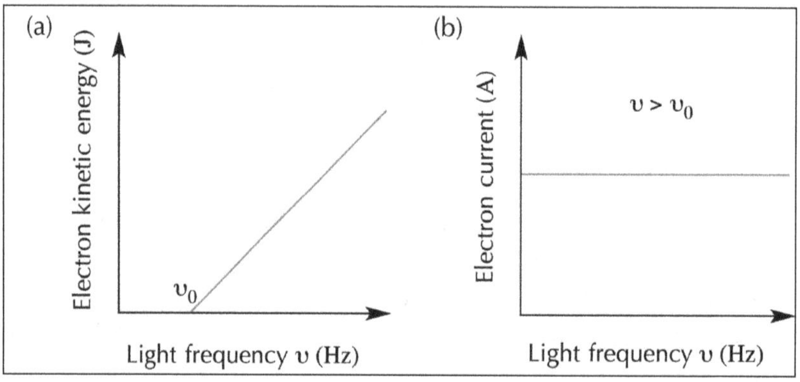

Because the light amplitude was kept constant as the light frequency increased, the number of photons being absorbed by the metal remained constant. Thus, the rate at which electrons were ejected from the metal (or the electric current) remained constant as well. The relationship between electron current and light frequency is illustrated in graph (b) above.

Isn't there more Math Somewhere?

We can analyze the frequency relationship using the law of conservation of energy. The total energy of the incoming photon, E_{photon}, must be equal to the kinetic energy of the ejected electron, $\text{KE}_{\text{electron}}$, plus the energy required to eject the electron from the metal. The energy required to free the electron from a particular metal is also called the metal's *work function*, which is represented by the symbol Φ (in units of J):

$$\text{E}_{\text{photon}} = \text{KE}_{\text{electron}} + \Phi$$

Like the threshold frequency v_0, the value of Φ also changes depending on the metal. We can now write the energy of the photon in terms of the light frequency using Planck's equation:

$$\text{E}_{\text{photon}} = hv = \text{KE}_{\text{electron}} + \Phi$$

Rearranging this equation in terms of the electron's kinetic energy, we get:

$$\text{KE}_{\text{electron}} = hv - \Phi$$

We can see that kinetic energy of the photoelectron increases linearly with ν as long as the photon energy is greater than the work function Φ, which is exactly the relationship shown in graph (a) above. We can also use this equation to find the photoelectron velocity v, which is related to $KE_{electron}$ as follows:

$$KE_{electron} = h\nu - \Phi = \frac{1}{2}m_e v^2$$

Where m_e is the rest mass of an electron, 9.1094×10^{-31} kg.

Exploring the Wave Amplitude Trends

In terms of photons, higher amplitude light means more photons hitting the metal surface. This results in more electrons ejected over a given time period. As long as the light frequency is greater than ν_0, increasing the light amplitude will cause the electron current to increase proportionally as shown in graph (a) below.

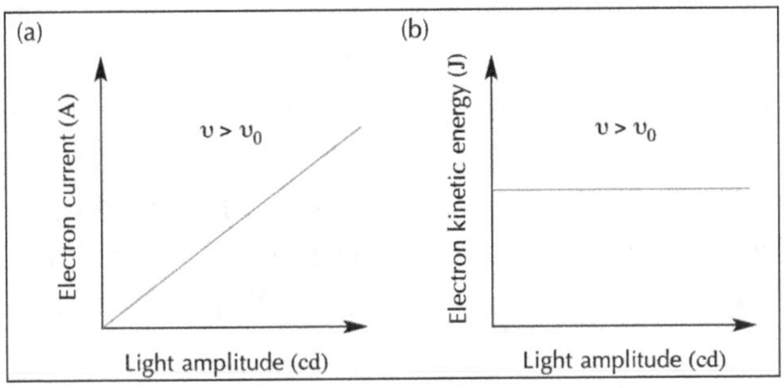

Since increasing the light amplitude has no effect on the energy of the incoming photon, the photoelectron kinetic energy remains constant as the light amplitude is increased.

If we try to explain this result using our dock-and-beach-balls analogy, the relationship in graph (b) indicates that no matter the height of the wave hitting the dock-whether it's a tiny swell, or a huge tsunami-the individual beach balls would be launched off the dock with the exact same speed! Thus, our intuition and analogy don't do a very good job of explaining these particular experiments.

The Compton Effect

The Compton effect is the term used for an unusual result observed when X-rays are scattered on some materials. By classical theory, when an electromagnetic wave is scattered off atoms, the wavelength of the scattered radiation is expected to be the same as the wavelength of the incident radiation. Contrary to this prediction of classical physics, observations show that when X-rays are scattered off some materials, such as graphite, the scattered X-rays have different wavelengths from the wavelength of the incident X-rays. This classically unexplainable phenomenon was studied experimentally by Arthur H. Compton and his collaborators, and Compton gave its explanation in 1923.

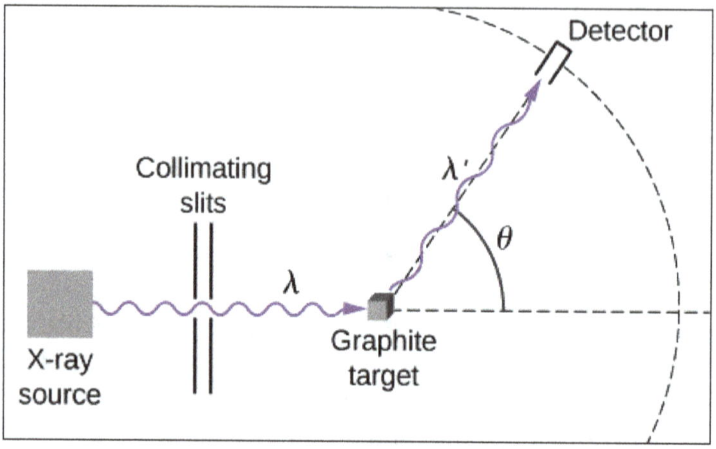

Experimental setup for studying Compton scattering.

To explain the shift in wavelengths measured in the experiment, Compton used Einstein's idea of light as a particle. The Compton effect has a very important place in the history of physics because it shows that electromagnetic radiation cannot be explained as a purely wave phenomenon. The explanation of the Compton effect gave a convincing argument to the physics community that electromagnetic waves can indeed behave like a stream of photons, which placed the concept of a photon on firm ground.

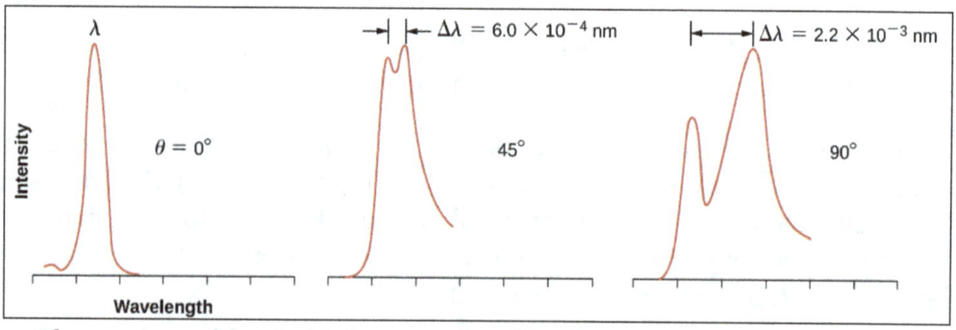

The experimental data in this figure are plotted in arbitrary units so that the height of the profile reflects the intensity of the scattered beam above background noise.

The schematics of Compton's experimental setup are shown in figure. The idea of the experiment is straightforward: Monochromatic X-rays with wavelength λ are incident on a sample of graphite (the "target"), where they interact with atoms inside the sample; they later emerge as scattered X-rays with wavelength λ′ . A detector placed behind the target can measure the intensity of radiation scattered in any direction θ with respect to the direction of the incident X-ray beam. This scattering angle, θ, is the angle between the direction of the scattered beam and the direction of the incident beam. In this experiment, we know the intensity and the wavelength λ of the incoming (incident) beam; and for a given scattering angle θ, we measure the intensity and the wavelength λ' of the outgoing (scattered) beam. Typical results of these measurements are shown in Figure, where the x-axis is the wavelength of the scattered X-rays and the y-axis is the intensity of the scattered X-rays, measured for different scattering angles (indicated on the graphs). For all scattering angles (except for θ=0°), we measure two intensity peaks. One peak is located at the wavelength λ, which is the wavelength of the incident beam. The other peak is located at some other wavelength, λ'. The two peaks are separated by Δλ, which depends on the scattering angle θ of the outgoing beam (in the direction of observation). The separation Δλ is called the Compton shift.

Compton Shift

As given by Compton, the explanation of the Compton shift is that in the target material, graphite, valence electrons are loosely bound in the atoms and behave like free electrons. Compton assumed that the incident X-ray radiation is a stream of photons. An incoming photon in this stream collides with a valence electron in the graphite target. In the course of this collision, the incoming photon transfers some part of its energy and momentum to the target electron and leaves the scene as a scattered photon. This model explains in qualitative terms why the scattered radiation has a longer wavelength than the incident radiation. Put simply, a photon that has lost some of its energy emerges as a photon with a lower frequency, or equivalently, with a longer wavelength. To show that his model was correct, Compton used it to derive the expression for the Compton shift. In his derivation, he assumed that both photon and electron are relativistic particles and that the collision obeys two commonsense principles:

- The conservation of linear momentum,
- The conservation of total relativistic energy.

In the following derivation of the Compton shift, E_f and $\vec{p}_f$ denote the energy and momentum, respectively, of an incident photon with frequency f. The photon collides with a relativistic electron at rest, which means that immediately before the collision, the electron's energy is entirely its rest mass energy, m_0c^2. Immediately after the collision, the electron has energy E and momentum $\vec{p}$, both of which satisfy equation $E_f = hf = \frac{hc}{\lambda}$. Immediately after the collision, the outgoing photon has energy $\tilde{E}_f$, momentum $\vec{\tilde{p}}_f$, and frequency f'. The direction of the incident photon is horizontal from left to right, and the direction of the outgoing photon is at the angle θ, as illustrated in figure. The scattering angle θ is the angle between the momentum vectors $\vec{p}_f$ and $\vec{\tilde{p}}_f$, and we can write their scalar product:

$$\vec{p} \cdot \vec{\tilde{p}}_f = p_f \vec{p}_f \cos\theta.$$

Following Compton's argument, we assume that the colliding photon and electron form an isolated system. This assumption is valid for weakly bound electrons that, to a good approximation, can be treated as free particles. Our first equation is the conservation of energy for the photon-electron system:

$$E_f + m_0 c^2 = \tilde{E}_f + E.$$

The left side of this equation is the energy of the system at the instant immediately before the collision, and the right side of the equation is the energy of the system at the instant immediately after the collision. Our second equation is the conservation of linear momentum for the photon–electron system where the electron is at rest at the instant immediately before the collision:

$$\vec{p}_f = \vec{\tilde{p}}_f + \vec{p}.$$

The left side of this equation is the momentum of the system right before the collision, and the right side of the equation is the momentum of the system right after collision. The entire physics of Compton scattering is contained in these three preceding equations––the remaining part is algebra. At this point, we could jump to the concluding formula for the Compton shift, but it is beneficial to highlight the main algebraic steps that lead to Compton's formula, which we give here as follows.

We start with rearranging the terms in Equation $E_f + m_0 c^2 = \tilde{E}_f + E$. and squaring it:

$$[(E_f - \tilde{E}_f) + m_0 c^2]^2 = E^2.$$

In the next step, we substitute Equation $E_f = hf = \frac{hc}{\lambda}$ for E^2, simplify, and divide both sides by c^2 to obtain:

$$(E_f / c - \tilde{E}_f / c)^2 + 2m_0 c(E_f / c - \tilde{E}_f / c) = p^2$$

Now we can use Equation $p_f = \frac{h}{\lambda}$ to express this form of the energy equation in terms of momenta. The result is:

$$(p_f - \tilde{p}_f)^2 + 2m_0 c(p_f - \tilde{p}_f) = p^2.$$

To eliminate p^2, we turn to the momentum equation Equation $\vec{p}_f = \vec{\tilde{p}}_f + \vec{p}$, rearrange its terms, and square it to obtain:

$$\begin{aligned}(\vec{p}_f - \vec{\tilde{p}}_f)^2 &= p^2 \\ &= p_f^2 + \tilde{p}_f^2 - 2p_f \tilde{p}_f \cos\theta\end{aligned}$$

The product of the momentum vectors is given by Equation $\vec{p}\cdot\vec{\tilde{p}}_f = p_f\tilde{p}_f \cos\theta$. When we substitute this result for p^2 in equation $(p_f-\tilde{p}_f)^2+2m_0c(p_f-\tilde{p}_f)=p^2$., we obtain the energy equation that contains the scattering angle θ:

$$(p_f-\tilde{p}_f)^2+2m_0c(p_f-\tilde{p}_f)=p_f^2+\tilde{p}_f^2-2p_f\tilde{p}_f\cos\theta.$$

With further algebra, this result can be simplified to:

$$\frac{1}{\tilde{p}_f}-\frac{1}{p_f}=\frac{1}{m_0c}(1-\cos\theta).$$

Now recall equation $p_f=\frac{h}{\lambda}$ and write: $1/\tilde{p}_f=\lambda'/h$ and $1/p_f=\lambda/h$. When these relations are substituted into equation $\frac{1}{\tilde{p}_f}-\frac{1}{p_f}=\frac{1}{m_0c}(1-\cos\theta)$, we obtain the relation for the Compton shift:

$$\lambda'-\lambda=\frac{h}{m_0c}(1-\cos\theta)$$

The factor h/m_0c is called the Compton wavelength of the electron:

$$\lambda_c=\frac{h}{m_0c}=0.00243\ nm=2.43\ pm$$

Denoting the shift as $\Delta\lambda=\lambda'-\lambda$, the concluding result can be rewritten as:

$$\Delta\lambda=\lambda_c(1-\cos\theta).$$

This formula for the Compton shift describes outstandingly well the experimental results shown in figure. Scattering data measured for molybdenum, graphite, calcite, and many other target materials are in accord with this theoretical result. The nonshifted peak shown in figure is due to photon collisions with tightly bound inner electrons in the target material. Photons that collide with the inner electrons of the target atoms in fact collide with the entire atom. In this extreme case, the rest mass in equation $\lambda_c=\frac{h}{m_0c}=0.00243\ nm=2.43\ pm$ must be changed to the rest mass of the atom. This type of shift is four orders of magnitude smaller than the shift caused by collisions with electrons and is so small that it can be neglected.

Compton scattering is an example of inelastic scattering, in which the scattered radiation has a longer wavelength than the wavelength of the incident radiation. In today's usage, the term "Compton scattering" is used for the inelastic scattering of photons by free, charged particles. In Compton scattering, treating photons as particles with momenta that can be transferred to charged particles provides the theoretical background to explain the wavelength shifts measured in experiments; this is the evidence that radiation consists of photons.

6
The Optical Fiber

Optical fiber is a transparent fiber made from silica and plastic that is widely used in fiber-optic communications. It can be categorized into single-mode optical fiber and multi-mode optical fiber. This chapter has been carefully written to provide an easy understanding of these types of optical fibers.

Fiber optics works a third way. It sends information coded in a beam of light down a glass or plastic pipe. It was originally developed for endoscopes in the 1950s to help doctors see inside the human body without having to cut it open first. In the 1960s, engineers found a way of using the same technology to transmit telephone calls at the speed of light (normally that's 186,000 miles or 300,000 km per second in a vacuum, but slows to about two thirds this speed in a fiber-optic cable).

Optical Technology

A section of 144-strand fiber-optic cable. Each strand is made of optically pure glass and is thinner than a human hair.

A fiber-optic cable is made up of incredibly thin strands of glass or plastic known as optical fibers; one cable can have as few as two strands or as many as several hundred. Each strand is less than a tenth as thick as a human hair and can carry something like 25,000 telephone calls, so an entire fiber-optic cable can easily carry several million calls.

Fiber-optic cables carry information between two places using entirely optical (light-based) technology. Suppose you wanted to send information from your computer to a friend's house down the street using fiber optics. You could hook your computer up to a laser, which would convert electrical information from the computer into a series of light pulses. Then you'd fire the laser down the fiber-optic cable. After traveling down the cable, the light beams would emerge at the other end. Your friend would need a photoelectric cell (light-detecting component) to turn the pulses of light back into electrical information his or her computer could understand. So the whole apparatus would be like a really neat, hi-tech version of the kind of telephone you can make out of two baked-bean cans and a length of string.

How Fiber-optics Works

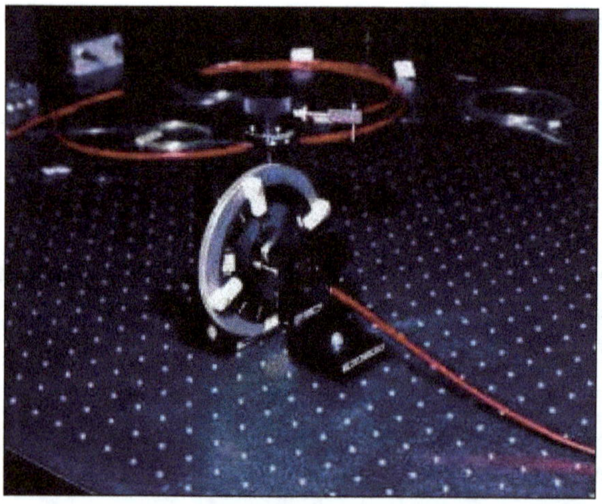

Fiber-optic cables are thin enough to bend, taking the light signals inside in curved paths too.

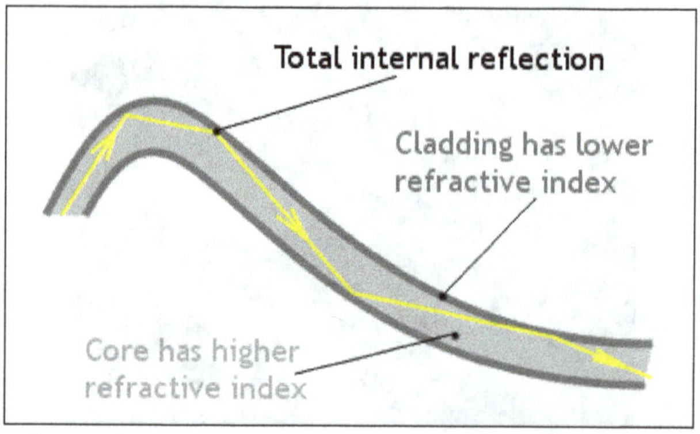

Total internal reflection keeps light rays bouncing down the inside of a fiber-optic cable.

Light travels down a fiber-optic cable by bouncing repeatedly off the walls. Each tiny photon (particle of light) bounces down the pipe like a bobsleigh going down an ice run. Now you might expect a beam of light, traveling in a clear glass pipe, simply to leak out of the edges. But if light hits glass at a really shallow angle (less than 42 degrees), it reflects back in again—as though the glass were really a mirror. This phenomenon is called total internal reflection. It's one of the things that keeps light inside the pipe.

The other thing that keeps light in the pipe is the structure of the cable, which is made up of two separate parts. The main part of the cable—in the middle—is called the core and that's the bit the light travels through. Wrapped around the outside of the core is another layer of glass called the cladding. The cladding's job is to keep the light signals inside the core. It can do this because it is made of a different type of glass to the core. (More technically, the cladding has a lower refractive index.)

Types of Fiber-optic Cables

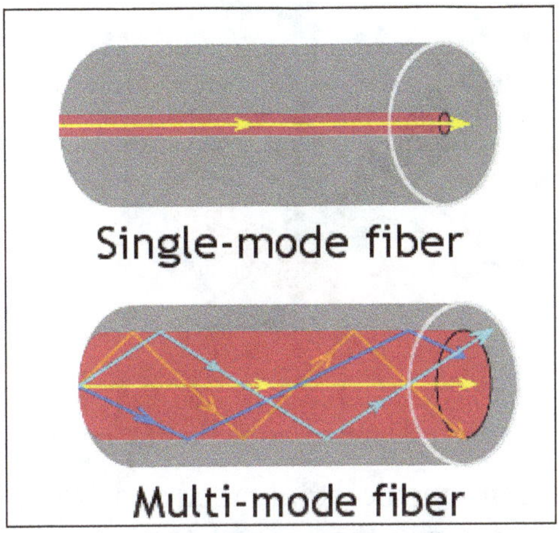

Optical fibers carry light signals down them in what are called modes. That sounds technical but it just means different ways of traveling: a mode is simply the path that a light beam follows down the fiber. One mode is to go straight down the middle of the fiber. Another is to bounce down the fiber at a shallow angle. Other modes involve bouncing down the fiber at other angles, more or less steep.

Artworks: Above: Light travels in different ways in single-mode and multi-mode fibers. Below: Inside a typical single-mode fiber cable (not drawn to scale). The thin core is surrounded by cladding roughly ten times bigger in diameter, a plastic outer coating (about twice the diameter of the cladding), some strengthening fibers made of a tough material such as Kevlar, with a protective outer jacket on the outside.

The simplest type of optical fiber is called single-mode. It has a very thin core about 5-10 microns (millionths of a meter) in diameter. In a single-mode fiber, all signals travel straight down the middle without bouncing off the edges (yellow line in diagram). Cable TV, Internet, and telephone signals are generally carried by single-mode fibers, wrapped together into a huge bundle. Cables like this can send information over 100 km (60 miles).

Another type of fiber-optic cable is called multi-mode. Each optical fiber in a multi-mode cable is about 10 times bigger than one in a single-mode cable. This means light beams can travel through the core by following a variety of different paths (yellow, orange, blue, and cyan lines)—in other words, in multiple different modes. Multi-mode cables can send information only over relatively short distances and are used (among other things) to link computer networks together.

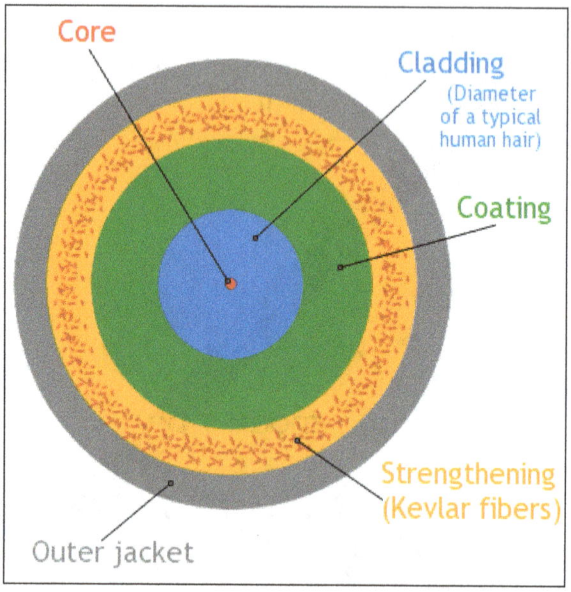

Even thicker fibers are used in a medical tool called a gastroscope (a type of endoscope), which doctors poke down someone's throat for detecting illnesses inside their stomach. A gastroscope is a thick fiber-optic cable consisting of many optical fibers. At the top end of a gastroscope, there is an eyepiece and a lamp. The lamp shines its light down one part of the cable into the patient's stomach. When the light reaches the stomach, it reflects off the stomach walls into a lens at the bottom of the cable. Then it travels back up another part of the cable into the doctor's eyepiece. Other types of endoscopes work the same way and can be used to inspect different parts of the body. There is also an industrial version of the tool, called a fiberscope, which can be used to examine things like inaccessible pieces of machinery in airplaneengines.

Uses for Fiber Optics

Shooting light down a pipe seems like a neat scientific party trick, and you might not think there'd be many practical applications for something like that. But just as electricity can power many types of machines, beams of light can carry many types of information—so they can help us in many ways. We don't notice just how commonplace fiber-optic cables have become because the laser-powered signals they carry flicker far beneath our feet, deep under office floors and city streets. The technologies that use it—computer networking, broadcasting, medical scanning, and military equipment (to name just four)—do so quite invisibly.

Working on fiber-optic cables.

Computer Networks

Fiber-optic cables are now the main way of carrying information over long distances because they have three very big advantages over old-style copper cables:

- Less attenuation: (signal loss) Information travels roughly 10 times further before it needs amplifying—which makes fiber networks simpler and cheaper to operate and maintain.
- No interference: Unlike with copper cables, there's no "crosstalk" (electromagnetic interference) between optical fibers, so they transmit information more reliably with better signal quality.
- Higher bandwidth: As we've already seen, fiber-optic cables can carry far more data than copper cables of the same diameter.

You're reading these words now thanks to the Internet. You probably chanced upon this page with a search engine like Google, which operates a worldwide network of giant data centers connected by vast-capacity fiber-optic cables (and is now trying to roll out fast fiber connections to the rest of us). Having clicked on a search engine link, you've downloaded this web page from my web server and my words have whistled most of the way to you down more fiber-optic cables. Indeed, if you're using fast fiber-optic broadband, optical fiber cables are doing almost all the work every time you go online. With most high-speed broadband connections, only the last part of the information's journey (the so-called "last mile" from the fiber-connected cabinet on your street to your house or apartment) involves old-fashioned wires. It's fiber-optic cables, not copper wires, that now carry "likes" and "tweets" under our streets, through an increasing number of rural areas, and even deep beneath the oceans linking continents. If you picture the Internet (and the World Wide Web that rides on it) as a global spider's web, the strands holding it together are fiber-optic cables; according to some estimates, fiber cables cover over 99 percent of the Internet's total mileage, and carry over 99 percent of all international communications traffic.

The faster people can access the Internet, the more they can—and will—do online. The arrival of broadband Internet made possible the phenomenon of cloud computing (where people store and process their data remotely, using online services instead of a home or business PC in their own premises). In much the same way, the steady rollout of fiber broadband (typically 5–10 times faster than conventional DSL broadband, which uses ordinary telephone lines) will make it much more commonplace for people to do things like streaming movies online instead of watching broadcast TV or renting DVDs. With more fiber capacity and faster connections, we'll be tracking and controlling many more aspects of our lives online using the so-called Internet of things.

Fiber-optic networks are expensive to construct (largely because it costs so much to dig up streets). Because the labor and construction costs are much more expensive than the cable itself, many network operators deliberately lay much more cable than they currently need.

But it's not just public Internet data that streams down fiber-optic lines. Computers were once connected over long distances by telephone lines or (over shorter distances) copper Ethernet cables, but fiber cables are increasingly the preferred method of networking computers because they're very affordable, secure, reliable, and have much higher capacity. Instead of linking its offices over the public Internet, it's perfectly possible for a company to set up its own fiber network (if it can afford to do so) or (more likely) buy space on a private fiber network. Many private computer networks run on what's called dark fiber, which sounds a bit sinister, but is simply the unused capacity on another network (optical fibers waiting to be lit up).

The Internet was cleverly designed to ferry any kind of information for any kind of use; it's not limited to carrying computer data. While telephone lines once carried the Internet, now the fiber-optic Internet carries telephone (and Skype) calls instead. Where telephone calls were once routed down an intricate patchwork of copper cables and microwave links between cities, most long-distance calls are now routed down fiber-optic lines. Vast quantities of fiber were laid from the 1980s onward; estimates vary wildly, but the worldwide total is believed to be several hundred million kilometers (enough to cross the United States about a million times). In the mid-2000s, it was estimated that as much as 98 percent of this was unused "dark fiber"; today, although much more fiber is in use, it's still generally believed that most networks contain anywhere from a third to a half dark fiber.

Broadcasting

Back in the early 20th century, radio and TV broadcasting was born from a relatively simple idea: it was technically quite easy to shoot electromagnetic waves through the air from a single transmitter (at the broadcasting station) to thousands of antennas on people's homes. These days, while radio still beams through the air, we're just as likely to get our TV through fiber-optic cables.

Cable TV companies pioneered the transition from the 1950s onward, originally using coaxial cables (copper cables with a sheath of metal screening wrapped around them to prevents crosstalk interference), which carried just a handful of analog TV signals. As more and more people connected to cable and the networks started to offer greater choice of channels and programs, cable operators found they needed to switch from coaxial cables to optical fibers and from analog to digitalbroadcasting. Fortunately, scientists were already figuring out how that might be possible; as far back as 1966, Charles Kao (and his colleague George Hockham) had done the math, proving how a single optical fiber cable might carry enough data for several hundred TV channels (or several hundred thousand telephone calls). It was only a matter of time before the world of cable TV took notice—and Kao's "groundbreaking achievement" was properly recognized when he was awarded the 2009 Nobel Prize in Physics.

Apart from offering much higher capacity, optical fibers suffer less from interference, so offer better signal (picture and sound) quality; they need less amplification to boost signals so they travel over long distances; and they're altogether more cost effective. In the future, fiber broadband may well be how most of us watch television, perhaps through systems such as IPTV (Internet Protocol Television), which uses the Internet's standard way of carrying data ("packet switching") to serve TV programs and movies on demand. While the copper telephone line is still the primary information route into many people's homes, in the future, our main connection to the world will be a high-bandwidth fiber-optic cable carrying any and every kind of information.

Medicine

Medical gadgets that could help doctors peer inside our bodies without cutting them open were the first proper application of fiber optics over a half century ago. Today, gastroscopes (as these things are called) are just as important as ever, but fiber optics continues to spawn important new forms of medical scanning and diagnosis.

One of the latest developments is called a lab on a fiber, and involves inserting hair-thin fiber-optic cables, with built-in sensors, into a patient's body. These sorts of fibers are similar in scale to the ones in communication cables and thinner than the relatively chunky light guides used in gastroscopes. How do they work? Light zaps through them from a lamp or laser, through the part of the body the doctor wants to study. As the light whistles through the fiber, the patient's body alters its properties in a particular way (altering the light's intensity or wavelength very slightly, perhaps). By measuring the way the light changes (using techniques such as interferometry), an instrument attached to the other end of the fiber can measure some critical aspect of how the patient's body is working, such as their temperature, blood pressure, cell pH, or the presence of medicines in their bloodstream. In other words, rather than simply using light to see inside the patient's body, this type of fiber-optic cable uses light to sense or measure it instead.

Military

Fiber optics on the battlefield. This Enhanced Fiber-Optic Guided Missile (EFOG-M) has an infrared fiber-optic camera mounted in its nose so that the gunner firing it can see where it's going as it travels.

It's easy to picture Internet users linked together by giant webs of fiber-optic cables; it's much less obvious that the world's hi-tech military forces are connected the same way. Fiber-optic cables are inexpensive, thin, lightweight, high-capacity, robust against attack, and extremely secure, so they offer perfect ways to connect military bases and other installations, such as missile launch sites and radar tracking stations. Since they don't carry electrical signals, they don't give off electromagnetic radiation that an enemy can detect, and they're robust against electromagnetic interference (including systematic enemy "jamming" attacks). Another benefit is the relatively light weight of fiber cables compared to traditional wires made of cumbersome and expensive copper metal. Tanks, military airplanes, and helicopters have all been slowly switching from metal cables to fiber-optic ones. Partly it's a matter of cutting costs and saving weight (fiber-optic cables weigh nearly 90 percent less than comparable "twisted-pair" copper cables). But it also improves reliability; for example, unlike traditional cables on an airplane, which have to be carefully shielded (insulated) to protect them against lightning strikes, optical fibers are completely immune to that kind of problem.

Type of Optical Fiber Modes

A mode is a stable propagation state in optical fibers. When light rays travel along certain paths through the optic fibers, the electromagnetic fields in the light waves support each other to form a stable field distribution. Thus light travels in the fibers. These stable operating points (standing waves) are called modes. If the light follows other paths then a stable wave will not propagate through the fiber and hence there will be no mode.

The optical fibers are typed according to the following modes:

Single Mode

In this, the light propagates in a single or fundamental mode in the core. Such fibers with only one mode are called single-mode fiber. It allows a single light path, and typically used with LASER signaling. The single mode fibers can allow greater bandwidth and cable runs than that of multimode but it is more expansive. The single mode fiber has the best characteristics of highest data rates and least attenuation. The single mode fiber is of very small size. It has the core of approximatly 5 to 10 micro meter in diameters.

Multi-mode

It is further divided into:

- Step-index.
- Graded-index.

Step-index Multimode Fibers

This fiber works in a very simplified way. The word step-index is used because the refractive index suddenly changes at the interface between core and cladding. The refractive index of the core is slightly greater than that of the cladding, thus confining the light to the core, by the principle of total internal reflection. The step-index multi mode fibers collect light easily but have a limited bandwidth.

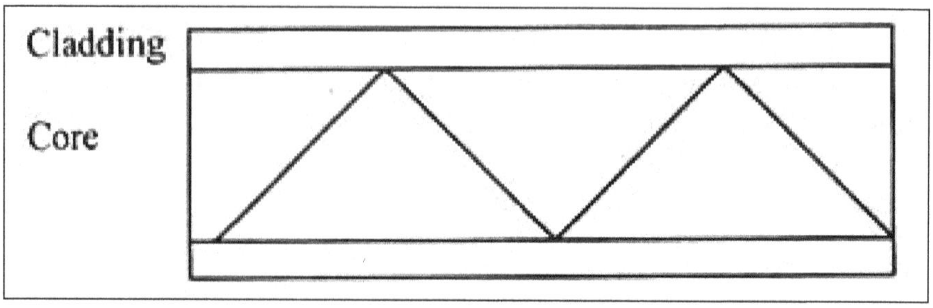

Step Index Multimode Fibers.

Graded-index Fibers

These are called graded-index fibers because in these fibers the refractive index changes gradually from the core to the cladding and at the boundary between the core and cladding, the change is abrupt. The refractive index decreases gradually from the center of the core to the edge of the cladding. Graded-index multi mode fibers collect light better than small core single mode fibers and have broader bandwidth than step-index multi mode fibers.

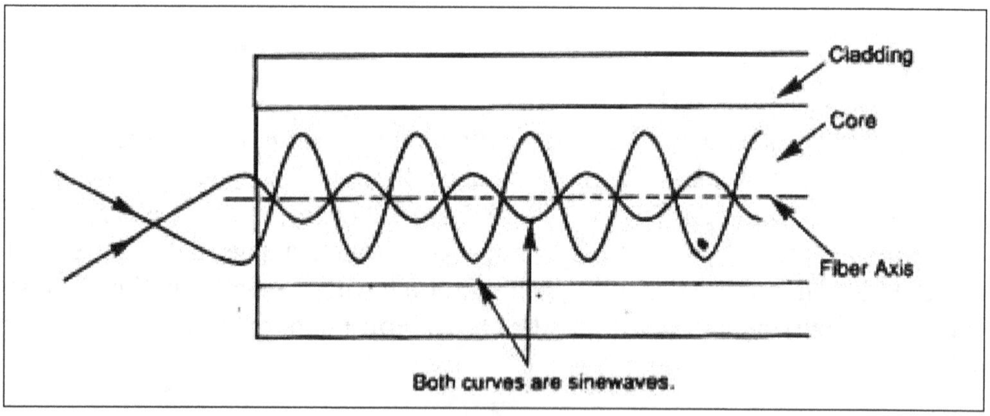

Graded Index Fibers.

Single-mode Optical Fiber

In fiber-optic communication, a single-mode optical fiber (SMF) is an optical fiber designed to carry light only directly down the fiber - the transverse mode. Modes are the possible solutions of the Helmholtz equation for waves, which is obtained by combining Maxwell's equations and the boundary conditions. These modes define the way the wave travels through space, i.e. how the wave is distributed in space. Waves can have the same mode but have different frequencies. This is the case in single-mode fibers, where we can have waves with different frequencies, but of the same mode, which means that they are distributed in space in the same way, and that gives us a single ray of light. Although the ray travels parallel to the length of the fiber, it is often called transverse mode since its electromagnetic oscillations occur perpendicular (transverse) to the length of the fiber. The 2009 Nobel Prize in Physics was awarded to Charles K. Kao for his theoretical work on the single-mode optical fiber. The standard G.652 defines the most widely used form of single-mode optical fiber.

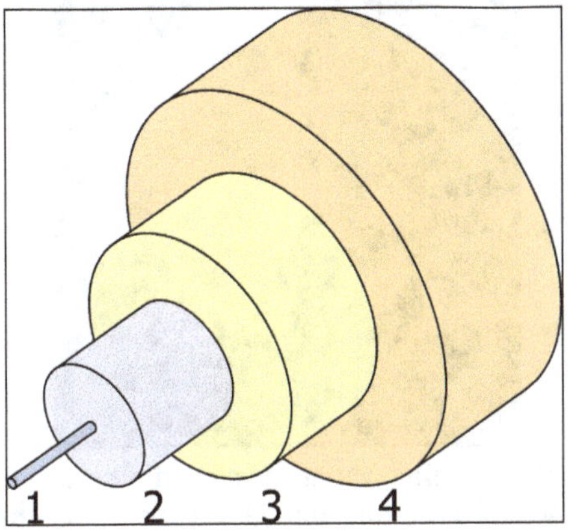

The structure of a typical single-mode fiber: 1. Core 8 - 9 µm diameter, 2. Cladding 125 µm dia, 3. Buffer 250 µm dia, 4. Jacket 900 µm dia.

Characteristics

Like multi-mode optical fibers, single-mode fibers do exhibit modal dispersion resulting from multiple spatial modes but with narrower modal dispersion. Single-mode fibers are therefore better at retaining the fidelity of each light pulse over longer distances than multi-mode fibers. For these reasons, single-mode fibers can have a higher bandwidth than multi-mode fibers. Equipment for single-mode fiber is more expensive than equipment for multi-mode optical fiber, but the single-mode fiber itself is usually cheaper in bulk.

A typical single-mode optical fiber has a core diameter between 8 and 10.5 µm and a cladding diameter of 125 µm. There are a number of special types of single-mode optical fiber which have been chemically or physically altered to give special properties, such as dispersion-shifted fiber and nonzero dispersion-shifted fiber. Data rates are limited by polarization mode dispersion and chromatic dispersion. As of 2005, data rates of up to 10 gigabits per second were possible at distances of over 80 km (50 mi) with commercially available transceivers (Xenpak). By using optical amplifiers and dispersion-compensating devices, state-of-the-art DWDM optical systems can span thousands of kilometers at 10 Gbit/s, and several hundred kilometers at 40 Gbit/s.

The lowest-order bounds mode is ascertained for the wavelength of interest by solving Maxwell's equations for the boundary conditions imposed by the fiber, which are determined by the core diameter and the refractive indices of the core and cladding. The solution of Maxwell's equations for the lowest order bound mode will permit a pair of orthogonally polarized fields in the fiber, and this is the usual case in a communication fiber.

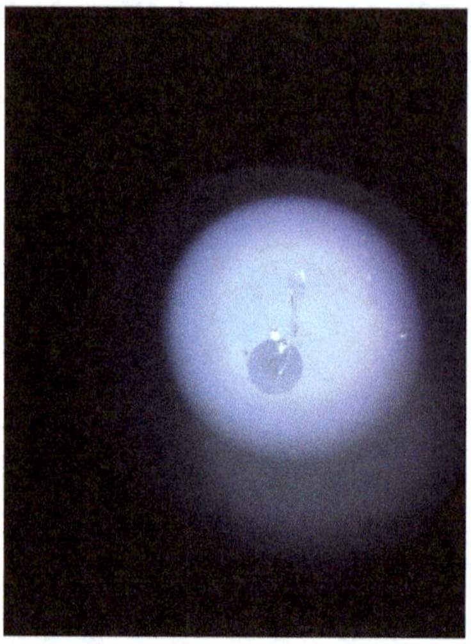

The round circle is the cladding, 125 microns in diameter. Debris is visible as a streak on the cross-section, and glows due to the illumination.

In step-index guides, single-mode operation occurs when the normalized frequency, V, is less than or equal to 2.405. For power-law profiles, single-mode operation occurs for a normalized frequency, V, less than approximately:

$$2.405\sqrt{\frac{g+2}{g}},$$

where g is the profile parameter.

In practice, the orthogonal polarizations may not be associated with degenerate modes.

OS1 and OS2 are standard single-mode optical fiber used with wavelengths 1310 nm and 1550 nm (size 9/125 µm) with a maximum attenuation of 1 dB/km (OS1) and 0.4 dB/km (OS2). OS1 is defined in ISO/IEC 11801, and OS2 is defined in ISO/IEC 24702.

Connectors

Optical fiber connectors are used to join optical fibers where a connect/disconnect capability is required. The basic connector unit is a connector assembly. A connector assembly consists of an adapter and two connector plugs. Due to the sophisticated polishing and tuning procedures that may be incorporated into optical connector manufacturing, connectors are generally assembled onto optical fiber in a supplier's manufacturing facility. However, the assembly and polishing operations involved can be performed in the field, for example to make cross-connect jumpers to size.

Optical fiber connectors are used in telephone company central offices, at installations on customer premises, and in outside plant applications. Their uses include:

- Making the connection between equipment and the telephone plant in the central office.
- Connecting fibers to remote and outside plant electronics such as Optical Network Units (ONUs) and Digital Loop Carrier (DLC) systems.
- Optical cross connects in the central office.
- Patching panels in the outside plant to provide architectural flexibility and to interconnect fibers belonging to different service providers.
- Connecting couplers, splitters, and Wavelength Division Multiplexers (WDMs) to optical fibers.
- Connecting optical test equipment to fibers for testing and maintenance.

Outside plant applications may involve locating connectors underground in subsurface enclosures that may be subject to flooding, on outdoor walls, or on utility poles. The closures that enclose them may be hermetic, or may be "free-breathing." Hermetic closures will prevent the connectors within being subjected to temperature swings unless they are breached. Free-breathing enclosures will subject them to temperature and humidity swings, and possibly to condensation and biological action from airborne bacteria, insects, etc. Connectors in the underground plant may be subjected to groundwater immersion if the closures containing them are breached or improperly assembled.

The latest industry requirements for optical fiber connectors are in Telcordia GR-326, Generic Requirements for Singlemode Optical Connectors and Jumper Assemblies.

A multi-fiber optical connector is designed to simultaneously join multiple optical fibers together, with each optical fiber being joined to only one other optical fiber.

The last part of the definition is included so as not to confuse multi-fiber connectors with a branching component, such as a coupler. The latter joins one optical fiber to two or more other optical fibers.

Multi-fiber optical connectors are designed to be used wherever quick and/or repetitive connects and disconnects of a group of fibers are needed. Applications include telecommunications companies' Central Offices (COs), installations on customer premises, and Outside Plant (OSP) applications.

The multi-fiber optical connector can be used in the creation of a low-cost switch for use in fiber optical testing. Another application is in cables delivered to a user with pre-terminated multi-fiber jumpers. This would reduce the need for field splicing, which could greatly reduce the number of hours necessary for placing an optical fiber cable in a telecommunications network. This, in turn, would result in savings for the installer of such cable.

Industry requirements for multi-fiber optical connectors are covered in GR-1435, Generic Requirements for Multi-Fiber Optical Connectors.

Fiber Optic Switches

An optical switch is a component with two or more ports that selectively transmits, redirects, or blocks an optical signal in a transmission medium. According to Telcordia GR-1073, an optical switch must be actuated to select or change between states. The actuating signal (also referred to as the control signal) is usually electrical, but in principle, could be optical or mechanical. (The control signal format may be Boolean and may be an independent signal; or, in the case of optical actuation, the control signal may be encoded in the input data signal. Switch performance is generally intended to be independent of wavelength within the component passband.)

Advantages

- No degradation of signal.
- Low dispersion.
- Well suited for long distance communication.

Disadvantages

- Manufacturing and handling is more difficult.
- Higher price.
- Coupling light into the fiber is difficult.

Multi-mode Optical Fibre

Multi-mode optical fiber is a type of optical fiber mostly used for communication over short distances, such as within a building or on a campus. Typical multi-mode links have data rates of 10 Mbit/s to 10 Gbit/s over link lengths of up to 600 meters (2000 feet). Multi-mode fiber has a fairly large core diameter that enables multiple light modes to be propagated and limits the maximum length of a transmission link because of modal dispersion.

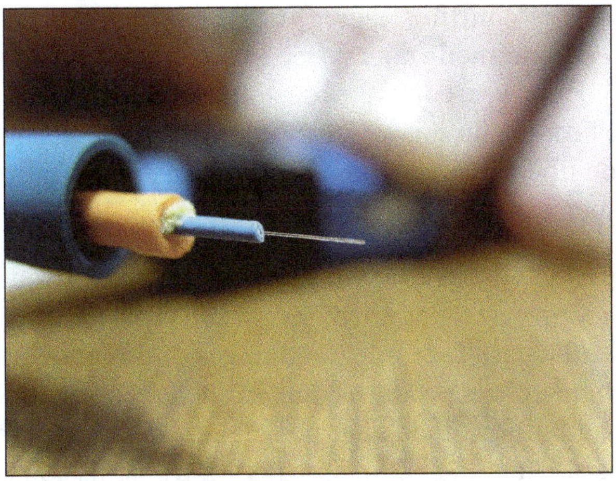

A stripped multi-mode fiber.

Applications

The equipment used for communications over multi-mode optical fiber is less expensive than that for single-mode optical fiber. Typical transmission speed and distance limits are 100 Mbit/s for distances up to 2 km (100BASE-FX), 1 Gbit/s up to 1000 m, and 10 Gbit/s up to 550 m.

Because of its high capacity and reliability, multi-mode optical fiber generally is used for backbone applications in buildings. An increasing number of users are taking the benefits of fiber closer to the user by running fiber to the desktop or to the zone. Standards-compliant architectures such as Centralized Cabling and fiber to the telecom enclosure offer users the ability to leverage the distance capabilities of fiber by centralizing electronics in telecommunications rooms, rather than having active electronics on each floor.

Multi-mode fiber is used for transporting light signals to and from miniature fiber optic spectroscopy equipment (spectrometers, sources, and sampling accessories) and was instrumental in the development of the first portable spectrometer.

Multi-mode fiber is also used when high optical powers are to be carried through an optical fiber, such as in laser welding.

Comparison with Single-mode Fiber

The main difference between multi-mode and single-mode optical fiber is that the former has much larger core diameter, typically 50–100 micrometers; much larger than the wavelength of the light carried in it. Because of the large core and also the possibility of large numerical aperture, multi-mode fiber has higher "light-gathering" capacity than single-mode fiber. In practical terms, the larger core size simplifies connections and also allows the use of lower-cost electronics such as light-emitting diodes (LEDs) and vertical-cavity surface-emitting lasers (VCSELs) which operate at the 850 nm and 1300 nm wavelength (single-mode fibers used in telecommunications typically operate at 1310 or 1550 nm). However, compared to single-mode fibers, the multi-mode fiber bandwidth–distance product limit is lower. Because multi-mode fiber has a larger core-size than single-mode fiber, it supports more than one propagation mode; hence it is limited by modal dispersion, while single mode is not.

The LED light sources sometimes used with multi-mode fiber produce a range of wavelengths and these each propagate at different speeds. This chromatic dispersion is another limit to the useful length for multi-mode fiber optic cable. In contrast, the lasers used to drive single-mode fibers produce coherent light of a single wavelength. Because of the modal dispersion, multi-mode fiber has higher pulse spreading rates than single mode fiber, limiting multi-mode fiber's information transmission capacity.

Single-mode fibers are often used in high-precision scientific research because restricting the light to only one propagation mode allows it to be focused to an intense, diffraction-limited spot.

Jacket color is sometimes used to distinguish multi-mode cables from single-mode ones. The standard TIA-598C recommends, for non-military applications, the use of a yellow jacket for single-mode fiber, and orange or aqua for multi-mode fiber, depending on type. Some vendors use violet to distinguish higher performance OM4 communications fiber from other types.

Types

Multi-mode fibers are described by their core and cladding diameters. Thus, 62.5/125 μm multi-mode fiber has a core size of 62.5 micrometres (μm) and a cladding diameter of 125 μm. The transition between the core and cladding can be sharp, which is called a step-index profile, or a gradual transition, which is called a graded-index profile. The two types have different dispersion characteristics and thus different effective propagation distance. Multi-mode fibers may be constructed with either graded or step-index profile.

In addition, multi-mode fibers are described using a system of classification determined by the ISO 11801 standard — OM1, OM2, and OM3 — which is based on the modal bandwidth of the multi-mode fiber. OM4 (defined in TIA-492-AAAD) was finalized in August 2009, and was published by the end of 2009 by the TIA. OM4 cable will support 125m links at 40 and 100 Gbit/s. The letters "OM" stand for optical multi-mode.

For many years 62.5/125 μm (OM1) and conventional 50/125 μm multi-mode fiber (OM2) were widely deployed in premises applications. These fibers easily support applications ranging from Ethernet (10 Mbit/s) to gigabit Ethernet (1 Gbit/s) and, because of their relatively large core size, were ideal for use with LED transmitters. Newer deployments often use laser-optimized 50/125 µm multi-mode fiber (OM3). Fibers that meet this designation provide sufficient bandwidth to support 10 Gigabit Ethernet up to 300 meters. Optical fiber manufacturers have greatly refined their manufacturing process since that standard was issued and cables can be made that support 10 GbE up to 400 meters. Laser optimized multi-mode fiber (LOMMF) is designed for use with 850 nm VCSELs.

The migration to LOMMF/OM3 has occurred as users upgrade to higher speed networks. LEDs have a maximum modulation rate of 622 Mbit/s because they cannot be turned on/off fast enough to support higher bandwidth applications. VCSELs are capable of modulation over 10 Gbit/s and are used in many high speed networks.

Some 200 and 400 Gigabit Ethernet speeds use wavelength-division multiplexing (WDM) even for multi-mode fiber which isn't specified up to and including OM4. In 2017, OM5 has been standardized by TIA and ISO for WDM MMF, specifying not only a minimum modal bandwidth for 850 nm but a curve spanning from 850 to 953 nm.

Cables can sometimes be distinguished by jacket color: for 62.5/125 µm (OM1) and 50/125 µm (OM2), orange jackets are recommended, while aqua is recommended for 50/125 µm "laser optimized" OM3 and OM4 fiber. Some fiber vendors use violet for "OM4+". OM5 is officially colored lime green.

VCSEL power profiles, along with variations in fiber uniformity, can cause modal dispersion which is measured by differential modal delay (DMD). Modal dispersion is caused by the different speeds of the individual modes in a light pulse. The net effect causes the light pulse to spread over distance, introducing intersymbol interference. The greater the length, the greater the modal dispersion. To combat modal dispersion, LOMMF is manufactured in a way that eliminates variations in the fiber which could affect the speed that a light pulse can travel. The refractive index profile is enhanced for VCSEL transmission and to prevent pulse spreading. As a result, the fibers maintain signal integrity over longer distances, thereby maximizing the bandwidth.

Comparison

Minimum reach of Ethernet variants over multi-mode fiber								
Category	Minimum modal bandwidth 850 / 953 / 1300 nm	Fast Ethernet 100 BASE-FX	1 Gb (1000 Mb) Ethernet 1000 BASE-SX	1 Gb (1000 Mb) Ethernet 1000 BASE-LX	10 Gb Ethernet 10 GBASE-SR	40 Gb Ethernet 40 GBASE-SWDM4	40 Gb Ethernet 40 GBASE-SR4	100 Gb Ethernet 100 GBASE-SR10

<table>
<tr><td>FDDI (62.5/125)</td><td>160 / – / 500 MHz·km</td><td rowspan="6">2000 m</td><td>220 m</td><td rowspan="3">550 m (mode-conditioning patch cord required)</td><td>26 m</td><td>Not Supported</td><td>Not supported</td><td>Not supported</td></tr>
<tr><td>OM1 (62.5/125)</td><td>200 / – / 500 MHz·km</td><td>275 m</td><td>33 m</td><td>Not Supported</td><td>Not supported</td><td>Not supported</td></tr>
<tr><td>OM2 (50/125)</td><td>500 / – / 500 MHz·km</td><td rowspan="4">550 m</td><td>82 m</td><td>Not Supported</td><td>Not supported</td><td>Not supported</td></tr>
<tr><td>OM3 (50/125) *Laser Optimized*</td><td>1500 / – / 500 MHz·km</td><td rowspan="3">550 m (no mode-conditioning patch cord should be used)</td><td>300 m</td><td>240m
Duplex LC</td><td>100 m
(330 m QSFP+ eSR)</td><td>100 m</td></tr>
<tr><td>OM4 (50/125) *Laser Optimized*</td><td>3500 / – / 500 MHz·km</td><td rowspan="2">400 m</td><td rowspan="2">350m
Duplex LC</td><td rowspan="2">150 m
(550 m QSFP+ eSR)</td><td rowspan="2">150 m</td></tr>
<tr><td>OM5 (50/125) "Wide-band multi-mode" for short-wave WDM</td><td>3500 / 1850 / 500 MHz·km</td></tr>
</table>

OFL Over-filled Launch for 850/953 nm/EMB Effective Modal Bandwidth for 1310 nm.

Difference between Step Index Fiber and Graded Index Fiber

Step Index Fiber

1. The refractive index of the core is uniform throughout and undergoes on abrupt change at the core cladding boundary.

2. The diameter of the core is about 50-200μm in the case of multimode fiber and 10μm in the case of single mode fiber.

3. The path of light propagation is zig- zag in manner.

4. Attenuation is more for multimode step index fiber but for single mode it is very less.

Explanation

When a ray travels through the longer distances there will be some difference in reflected angles. Hence high angle rays arrive later than low angle rays causing dispersion resulting in distorted output.

4. This fiber has lower bandwidth.

5. The light ray propagation is in the form of meridional rays and it passes through the fiber axis.

Graded Index Fiber

1. The refractive index of the core is made to vary gradually such that it is maximum at the center of the core.

2. The diameter of the core is about 50μm in the case of multimode fiber.

3. The path of light is helical in manner.

4. Attenuation is less.

Explanation

Here the light rays travel with different velocity inn different paths because of their variation in their refractive indices. At the outer edge it travels faster than near the center. But almost all the rays reach the exit at the same time due to helical path. Thus, there is no dispersion.

1. This fiber has higher bandwidth.

2. The light propagation is in the form of skew rays and it will not cross fiber axis.

Propagation of Light in Grin Fiber

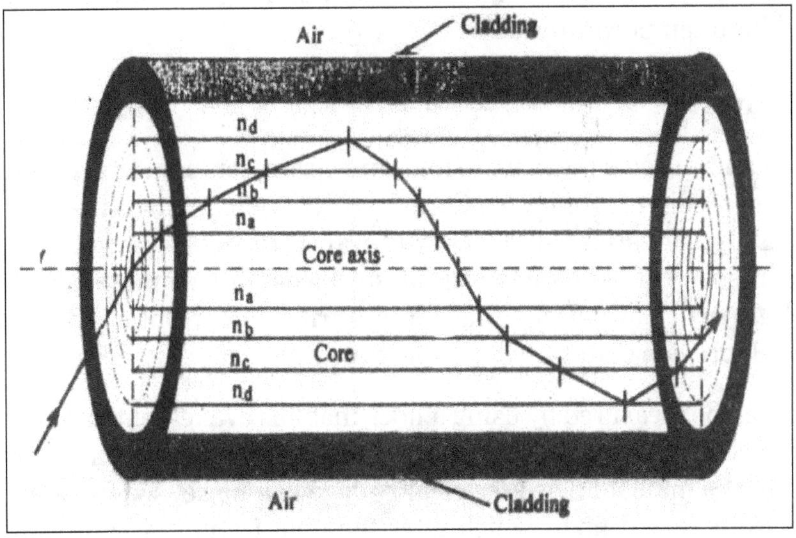

Let na, nb, nc, nd etc be the refractive index of different layers in graded index fiber with na > nb> nc> nd etc. then the propagation of light through the graded index fiber is as shown in the figure.

Here, since na > nb the ray gets refracted. Similarly since nb> nc, the ray gets refracted and so on. In a similar manner, due to decrease in refractive index the ray gets gradually curved towards the upward direction and at one place, where in it satisfies the condition for total internal reflection, $(\phi > \phi c)$ it is totally internally reflected.

The reflected rays travels back towards the core axis and without crossing the fiber axis, it is refracted towards downwards direction and again gets totally internally reflected and passes towards upward direction. In this manner the ray propagates inside the fiber in a helical or spiral manner.

Fiber Optic Sensors

A fiber-optic sensor is a sensor that uses optical fiber either as the sensing element ("intrinsic sensors"), or as a means of relaying signals from a remote sensor to the electronics that process the signals ("extrinsic sensors"). Fibers have many uses in remote sensing. Depending on the application, fiber may be used because of its small size, or because no electrical power is needed at the remote location, or because many sensors can be multiplexed along the length of a fiber by using light wavelength shift for each sensor, or by sensing the time delay as light passes along the fiber through each sensor. Time delay can be determined using a device such as an optical time-domain reflectometer and wavelength shift can be calculated using an instrument implementing optical frequency domain reflectometry.

Fiber-optic sensors are also immune to electromagnetic interference, and do not conduct electricity so they can be used in places where there is high voltage electricity or flammable material such as jet fuel. Fiber-optic sensors can be designed to withstand high temperatures as well.

Intrinsic Sensors

Optical fibers can be used as sensors to measure strain, temperature, pressure and other quantities by modifying a fiber so that the quantity to be measured modulates the intensity, phase, polarization, wavelength or transit time of light in the fiber. Sensors that vary the intensity of light are the simplest, since only a simple source and detector are required. A particularly useful feature of intrinsic fiber-optic sensors is that they can, if required, provide distributed sensing over very large distances.

Temperature can be measured by using a fiber that has evanescent loss that varies with temperature, or by analyzing the Rayleigh Scattering, Raman scattering or the Brillouin scattering in the optical fiber. Electrical voltage can be sensed by nonlinear optical effects in specially-doped fiber, which alter the polarization of light as a function of voltage or electric field. Angle measurement sensors can be based on the Sagnac effect.

Special fibers like long-period fiber grating (LPG) optical fibers can be used for direction recognition. Photonics Research Group of Aston University in UK has some publications on vectorial bend sensor applications.

Optical fibers are used as hydrophones for seismic and sonar applications. Hydrophone systems with more than one hundred sensors per fiber cable have been developed. Hydrophone sensor systems are used by the oil industry as well as a few countries' navies. Both bottom-mounted hydrophone arrays and towed streamer systems are in use. The German company Sennheiser developed a laser microphone for use with optical fibers.

A fiber-optic microphone and fiber-optic based headphone are useful in areas with strong electrical or magnetic fields, such as communication amongst the team of people working on a patient inside a magnetic resonance imaging (MRI) machine during MRI-guided surgery.

Optical fiber sensors for temperature and pressure have been developed for downhole measurement in oil wells. The fiber-optic sensor is well suited for this environment as it functions at temperatures too high for semiconductor sensors (distributed temperature sensing).

Optical fibers can be made into interferometric sensors such as fiber-optic gyroscopes, which are used in the Boeing 767 and in some car models (for navigation purposes). They are also used to make hydrogen sensors.

Fiber-optic sensors have been developed to measure co-located temperature and strain simultaneously with very high accuracy using fiber Bragg gratings. This is particularly useful when acquiring information from small or complex structures. Fiber optic sensors are also particularly well suited for remote monitoring, and they can be interrogated 290 km away from the monitoring station using an optical fiber cable. Brillouin scattering effects can also be used to detect strain and temperature over large distances (20–120 kilometers).

Other Examples

A fiber-optic AC/DC voltage sensor in the middle and high voltage range (100–2000 V) can be created by inducing measurable amounts of Kerr nonlinearity in single mode optical fiber by exposing a calculated length of fiber to the external electric field. The measurement technique is based on polarimetric detection and high accuracy is achieved in a hostile industrial environment.

High frequency (5 MHz–1 GHz) electromagnetic fields can be detected by induced nonlinear effects in fiber with a suitable structure. The fiber used is designed such that the Faraday and Kerr effects cause considerable phase change in the presence of the external field. With appropriate sensor design, this type of fiber can be used to measure different electrical and magnetic quantities and different internal parameters of fiber material.

Electrical power can be measured in a fiber by using a structured bulk fiber ampere sensor coupled with proper signal processing in a polarimetric detection scheme. Experiments have been carried out in support of the technique.

Fiber-optic sensors are used in electrical switchgear to transmit light from an electrical arc flash to a digital protective relay to enable fast tripping of a breaker to reduce the energy in the arc blast.

Fiber Bragg grating based fiber-optic sensors significantly enhance performance, efficiency and safety in several industries. With FBG integrated technology, sensors can provide detailed analysis and comprehensive reports on insights with very high resolution. These type of sensors are used extensively in several industries like telecommunication, automotive, aerospace, energy, etc. Fiber Bragg gratings are sensitive to the static pressure, mechanical tension and compression and fiber temperature changes. The efficiency of fiber Bragg grating based fiber-optic sensors can be provided by means of central wavelength adjustment of light emitting source in accordance with the current Bragg gratings reflection spectra.

Extrinsic Sensors

Extrinsic fiber-optic sensors use an optical fiber cable, normally a multimode one, to transmit modulated light from either a non-fiber optical sensor, or an electronic sensor connected to an optical transmitter. A major benefit of extrinsic sensors is their ability to reach places which are otherwise inaccessible. An example is the measurement of temperature inside aircraft jet engines by using a fiber to transmit radiation into a radiation pyrometer located outside the engine. Extrinsic sensors can also be used in the same way to measure the internal temperature of electrical transformers, where the extreme electromagnetic fields present make other measurement techniques impossible.

Extrinsic fiber-optic sensors provide excellent protection of measurement signals against noise corruption. Unfortunately, many conventional sensors produce electrical output which must be converted into an optical signal for use with fiber. For example, in the case of a platinum resistance thermometer, the temperature changes are translated into resistance changes. The PRT must therefore have an electrical power supply. The modulated voltage level at the output of the PRT can then be injected into the optical fiber via the usual type of transmitter. This complicates the measurement process and means that low-voltage power cables must be routed to the transducer.

Extrinsic sensors are used to measure vibration, rotation, displacement, velocity, acceleration, torque, and temperature.

Chemical Sensors and Biosensors

It is well-known the propagation of light in optical fiber is confined in the core of the fiber based on the total internal reflection (TIR) principle and near-zero propagation loss within the cladding, which is very important for the optical communication but limits its sensing applications due to the non-interaction of light with surroundings. Therefore, it is essential to exploit novel fiber-optic structures to disturb the light propagation, thereby enabling the interaction of the light with surroundings and constructing fiber-optic sensors. Until now, several methods, including polishing, chemical etching, tapering, bending, as well as femtosecond grating inscription, have been proposed to tailor the light propagation and prompt the interaction of light with sensing materials. In the above-mentioned fiber-optic structures, the enhanced evanescent fields can be efficiently excited to induce the light to expose to and interact with the surrounding medium. However, the fibers themselves can only sense very few kinds of analytes with low-sensitivity and zero-selectivity, which greatly limits their development and applications, especially for biosensors that require both high-sensitivity and high-selectivity. To overcome the issue, an efficient way is to resort to responsive materials, which possess the ability to change their properties, such as RI, absorption, conductivity, etc., once the surrounding environments change. Due to the rapid progress of functional materials in recent years, various sensing materials are available for fiber-optic chemical sensors and biosensors fabrication, including graphene, metals and metal oxides, carbon nanotubes, nanowires, nanoparticles, polymers, quantum dots, etc. Generally, these materials reversibly change their shape/volume upon stimulation by the surrounding environments (the target analysts), which then leads to the variation of RI or absorption of the sensing materials. Consequently, the surrounding changes will be recorded and interrogated by the optical fibers, realizing sensing functions of optical fibers. Currently, various fiber-optic chemical sensors and biosensors have been proposed and demonstrated.

Fiber Optics Communication

Fibre optic communication has revolutionised the telecommunications industry. It has also made its presence widely felt within the data networking community as well. Using fibre optic cable, optical communications have enabled telecommunications links to be made over much greater distances and with much lower levels of loss in the transmission medium and possibly most important of all, fiber optical communications has enabled much higher data rates to be accommodated.

As a result of these advantages, fibre optic communications systems are widely employed for applications ranging from major telecommunications backbone infrastructure to Ethernet systems, broadband distribution, and general data networking.

Development of Fibre Optics

Since the earliest days of telecommunications there has been an ever increasing need to transmit more data even faster. Initially single line wires were used. These gave way to coaxial cables that enabled several channels to transmitted over the same cable. However these systems were limited in bandwidth and optical systems were investigated.

Optical communications became a possibility after the first lasers were developed in the 1960s. The next piece of the jigsaw fell into place when the first optical fibers with a sufficiently low loss for communications purposes were developed in the 1970s. Then, during the late 1970s a considerable amount of research was undertaken. This resulted in the installation of the first optical fibre telecommunications system. It ran over a distance of 45 km and used a wavelength of 0.5 mm and had a data rate of just 45 Mbps - a fraction of what is possible today.

Since then, considerable improvements have been made in the technology. Data rates have improved and in addition to this the performance of the optical fibre has been improved to enable much greater distances to be achieved between repeaters. As an indication of this the speeds that can now be achieved along through a fibre optic system exceed 10 Tbps.

When the first fibre optic transmission systems were being developed, it was thought that the fibre optic cabling and technology would be prohibitively expensive. However, this has not been the case and costs have fallen to the extent that fibre optics now provides the only viable option for many telecommunications applications. In addition to this it is also used in many local area networks where speed is a major requirement.

Advantages of Fibre Optics for Communications

There are a number of compelling reasons that lead to the widespread adoption of fibre optic cabling for telecommunications applications:

- Much lower levels of signal attenuation.
- Fibre optic cabling provides a much higher bandwidth allowing more data to be delivered.
- Fibre optic cables are much lighter than the coaxial cables that might otherwise be used.
- Fibre optics do not suffer from stray interference pickup that occurs with coaxial cabling.

Fibre Optic Transmission System

Any fibre optic data transmission system will comprise a number of different elements.

There are three major elements (marked in bold), and a further one that is vital for practical systems:

- Transmitter (light source).
- Fibre optic cable.
- Optical repeater.
- Receiver (Detector).

The different elements of the system will vary according to the application. Systems used for lower capacity links, possibly for local area networks will employ somewhat different techniques and components to those used by network providers that provide extremely high data rates over long distances. Nevertheless the basic principles are the same whatever the system.

In the system the transmitter of light source generates a light stream modulated to enable it to carry the data. Conventionally a pulse of light indicates a "1" and the absence of light indicates "0". This light is transmitted down a very thin fibre of glass or other suitable material to be presented at the receiver or detector. The detector converts the pulses of light into equivalent electrical pulses. In this way the data can be transmitted as light over great distances.

Fibre Optic Transmitter

Although the original telecommunications fibre optic systems would have used large lasers, today a variety of semiconductor devices can be used. The most commonly used devices are light emitting diodes, LEDs, and semiconductor laser diodes.

The simplest transmitter device is the LED. Its main advantage is that it is cheap, and this makes it ideal for low cost applications where only short runs are needed. However they have a number of drawbacks. The first is that they offer a very low level of efficiency. Only about 1% of the input power enters the optical fibre, and this means that high power drivers would be needed to provide sufficient light to enable long distance transmissions to be made. The other disadvantage of LEDs is that they produce what is termed incoherent light that covers a relatively wide spectrum. Typically the spectral width is between 30 and 60 nm. This means that any chromatic dispersion in the fibre will limit the bandwidth of the system.

In view of their performance, LEDs are used mainly in local-area-network applications where the data rates are typically in the range 10-100 Mb/s and transmission distances are a few kilometres.

Where higher levels of performance are required, i.e. it is necessary that the fibre optic link can operate over greater distances and with higher data rates, then lasers are used. Although more costly, they offer some significant advantages. In the first instance they are able to provide a higher output level, and in addition to this the light output is directional and this enables a much higher level of efficiency in the transfer of the light into the fibre optic cable. Typically the coupling efficiency into a single mode fibre may be as high as 50%. A further advantage is that lasers have a very narrow spectral bandwidth as a result of the fact that they produce coherent light. This narrow spectral width enables the lasers to transmit data at much higher rates because modal dispersion is less apparent. Another advantage is that semiconductor lasers can be modulated directly at high frequencies because of short recombination time for the carriers within the semiconductor material.

Laser diodes are often directly modulated. This provides a very simple and effective method of transferring the data onto the optical signal. This is achieved by controlling current applied directly to the device. This in turn varies the light output from the laser. However for very high data rates or very long distance links, it is more effective to run the laser at a constant output level (continuous wave). The light is then modulated using an external device. The advantage of using an external means of modulation is that it increases the maximum link distance because an effect known as laser chirp is eliminated. This chirp broadens the spectrum of the light signal and this increases the chromatic dispersion in the fibre optic cable.

Fibre Optic Cable

In essence a fibre optic cable consists of core, around which is another layer referred to as the cladding. Outside of this there is a protective outer coating.

The fibre optic cables operate because their cladding has a refractive index that is slightly lower than that of the core. This means that light passing down the core undergoes total internal reflection when it reaches the core/cladding boundary, and it is thereby contained within the core of the optical fibre.

Repeaters and Amplifiers

There is a maximum distance over which signals may be transmitted over fibre optic cabling. This is limited not only by the attenuation of the cable, but also the distortion of the light signal along the cable. In order to overcome these effects and transmit the signals over longer distances, repeaters and amplifiers are used.

Opto-electric repeaters may be used. These devices convert the optical signal into an electrical format where it can be processed to ensure that the signal is not distorted and then converted back into the optical format. It may then be transmitted along the next state of the fibre optic cable.

An alternative approach is to use an optical amplifier. These amplifiers directly amplify the optical signal without the need to convert the signal back into an electrical format. The amplifiers consist of a length of fibre optic cable that is doped with a rare earth mineral named Erbium. The treated fibre cable is then illuminated or pumped with light of a shorter wavelength from another laser and this serves to amplify the signal that is being carried.

In view of the much reduced cost of fibre optic amplifiers over repeaters, amplifiers are far more widely used. Most repeaters have been replaced, and amplifiers are used in virtually all new installations these days.

Receivers

Light travelling along a fibre optic cable needs to be converted into an electrical signal so that it can be processed and the data that is carried can be extracted. The component that is at the heart of the receiver is a photo-detector. This is normally a semiconductor device and may be a p-n junction, a p-i-n photo-diode or an avalanche photo-diode. Photo-transistors are not used because they do not have sufficient speed.

Once the optical signal from the fibre optic cable has been applied to the photo-detector and converted into an electrical format it can be processed to recover the data which can then be passed to its final destination.

Fibre optic transmission of data is generally used for long distance telecommunications network links and for high speed local area networks. Currently fibre optics is not used for the delivery of services to homes, although this is a long term aim for many telcos. By using optical fibre cabling here, the available bandwidth for new services would be considerably higher and the possibility of greater revenues would increase. Currently the cost oft his is not viable, although it is likely to happen in the medium term.

References

- Fiberoptics: explainthatstuff.com, Retrieved 10 January, 2019
- Crawford, Dwayne (Sep 11, 2013). "Who is Erika Violet and what is she doing in my data center?". Tech Topics. Belden. Retrieved Feb 12, 2014
- Optical-fiber, communication-system, electronics: daenotes.com, Retrieved 11 February, 2019
- Measures, Raymond M. (2001). Structural Monitoring with Fiber Optic Technology. San Diego, California, USA: Academic Press. Pp. Chapter 7. ISBN 978-0-12-487430-5
- Difference-between-Step-Index-fiber-and-Graded-Index-fiber-6890: brainkart.com, Retrieved 12 March, 2019
- Ghosh, S.K.; Sarkar, S.K.; Chakraborty, S. (2006). "A proposal for single mode fiber optic watt measurement scheme". Journal of Optics (Calcutta). 35 (2): 118–124. Doi:10.1007/BF03354801. ISSN 0972-8821
- Optical-fibre-telecommunications-basics, fibre-optics, connectivity: electronics-notes.com, Retrieved 13 April, 2019

7
Understanding Lasers

The device that produces light through the process of optical amplification is known as a laser. Some of the types of lasers are solid-state lasers, liquid lasers, gas lasers and semiconductor lasers. This chapter discusses the characteristics and working of these types of lasers.

The word "laser" is an acronym for Light Amplification by Stimulated Emission of Radiation. Lasers are finding ever increasing military applications -- principally for target acquisition, fire control, and training. These lasers are termed rangefinders, target designators, and direct-fire simulators. Lasers are also being used in communications, laser radars (LIDAR), landing systems, laser pointers, guidance systems, scanners, metal working, photography, holography, and medicine.

In this document the word laser will be limited to electromagnetic radiation emitting devices using light amplification by stimulated emission of radiation at wavelengths from 180 nanometers to 1 millimeter. The electromagnetic spectrum includes energy ranging from gamma rays to electricity. Figure illustrates the total electromagnetic spectrum and wavelengths of the various regions.

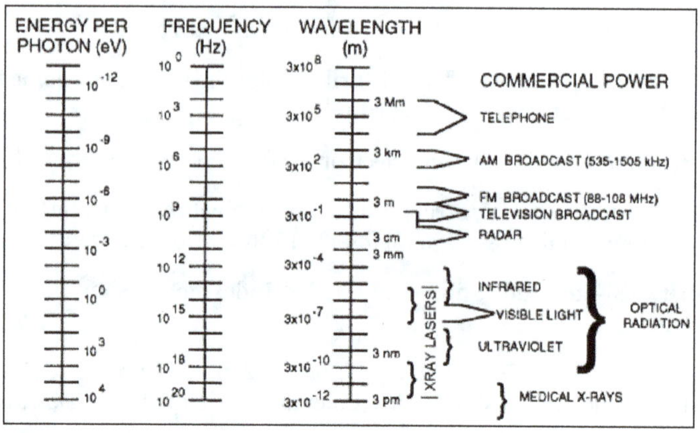

Electromagnetic Spectrum.

The primary wavelengths of laser radiation for current military and commercial applications include the ultraviolet, visible, and infrared regions of the spectrum. Ultraviolet radiation for lasers consists of wavelengths between 180 and 400 nm. The visible region consists of radiation with wavelengths between 400 and 700 nm. This is the portion we call visible light. The infrared region of the spectrum consists of radiation with wavelengths between 700 nm and 1 mm. Laser radiation absorbed by the skin penetrates only a few layers. In the eye, visible and near infrared radiation passes through the cornea, and is focused on and absorbed by the retina. It is the wavelength of the light that determines the visible sensation of color: violet at 400 nm, red at 700 nm, and the other colors of the visible spectrum in between. When radiation is absorbed, the effect on the absorbing biological tissue is either photochemical, thermal, or mechanical: in the ultraviolet region, the action is primarily photochemical; in the infrared region, the action is primarily thermal; and in the visible region, both effects are present. When the intensity of the radiation is sufficiently high, damage to the absorbing tissue will result.

Laser Theory and Operation

A basic understanding of how a laser operates helps in understanding the hazards when using a laser device. Figure shows that electromagnetic radiation is emitted whenever a charged particle such as an electron gives up energy. This happens every time an electron drops from a higher energy state, Q_1, to a lower energy state, Q_0, in an atom or ion as occurs in a fluorescent light. This also happens from changes in the vibrational or rotational state of molecules.

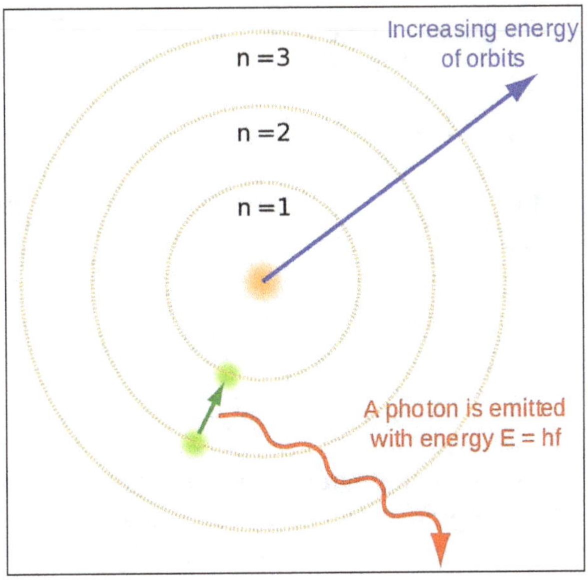

Emission of radiation from an atom by transition of an electron from a higher energy state to a lower energy state.

The color of light is determined by its frequency or wavelength. The shorter wavelengths are the ultraviolet and the longer wavelengths are the infrared. The smallest particle of light energy is described by quantum mechanics as a photon. The energy, E, of a photon is determined by its frequency, v, and Planck's constant, h.

$$E = h \cdot v$$

The velocity of light in a vacuum, c, is 300 million meters per second. The wavelength, λ, of light is related to v from the following equation:

$$\lambda = \frac{c}{v}$$

The difference in energy levels across which an excited electron drops determines the wavelength of the emitted light.

Components of A Laser

As shown in figure, the three basic components of a laser are:

- Lasing material (crystal, gas, semiconductor, dye, etc).
- Pump source (adds energy to the lasing material, e.g. flash lamp, electrical current to cause electron collisions, radiation from a laser, etc).
- Optical cavity consisting of reflectors to act as the feedback mechanism for light amplification.

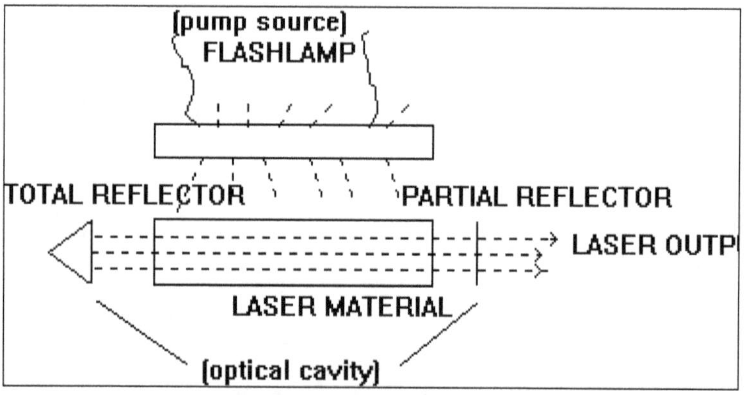

Solid State Laser Diagram.

Electrons in the atoms of the lasing material normally reside in a steady-state lower energy level. When light energy from the flashlamp is added to the atoms of the lasing material, the majority of the electrons are excited to a higher energy level - a phenomenon known as population inversion. This is an unstable condition for these electrons. They will stay in this state for a short time and then decay back to their original energy state. This decay occurs in two ways: spontaneous decay - the electrons simply fall to their ground state while emitting randomly directed photons; and stimulated decay

- the photons from spontaneous decaying electrons strike other excited electrons which causes them to fall to their ground state. This stimulated transition will release energy in the form of photons of light that travel in phase at the same wavelength and in the same direction as the incident photon. If the direction is parallel to the optical axis, the emitted photons travel back and forth in the optical cavity through the lasing material between the totally reflecting mirror and the partially reflecting mirror. The light energy is amplified in this manner until sufficient energy is built up for a burst of laser light to be transmitted through the partially reflecting mirror.

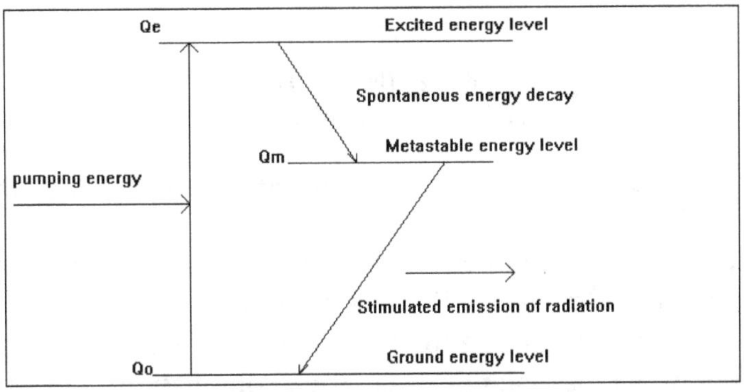

Three level laser energy diagram.

As shown in figure above, a lasing medium must have at least one excited (metastable) state where electrons can be trapped long enough (microseconds to milliseconds) for a population inversion to occur. Although laser action is possible with only two energy levels, most lasers have four or more levels.

A Q-switch in the optical path is a method of providing laser pulses of an extremely short time duration. A rotating prism like the total reflector in figure was an early method of providing Q-switching. Only at the point of rotation when there is a clear optical path will light energy be allowed to pass. A normally opaque electro-optical device (e.g., a pockels cell) is now often used for a Q-switching device. At the time of voltage application, the device becomes transparent, the light built up in the cavity by excited atoms can then reach the mirror so that the cavity Quality, Q, increases to a high level and emits a high peak power laser pulse of a few nanoseconds duration. When the phases of different frequency modes of a laser are synchronized (locked together), these modes will interfere with each other and generate a beat effect. The result is a laser output with regularly spaced pulsations called "mode locking". Mode locked lasers usually produce trains of pulses with a duration of a few picoseconds to nanoseconds resulting in higher peak powers than the same laser operating in the Q-switched mode. Pulsed lasers are often designed to produce repetitive pulses. The pulse repetition frequency, prf, as well as pulse width is extremely important in evaluating biological effects.

Types of Lasers

The laser diode is a light emitting diode with an optical cavity to amplify the light emitted from the energy band gap that exists in semiconductors as shown in figure. They can be tuned by varying the applied current, temperature or magnetic field.

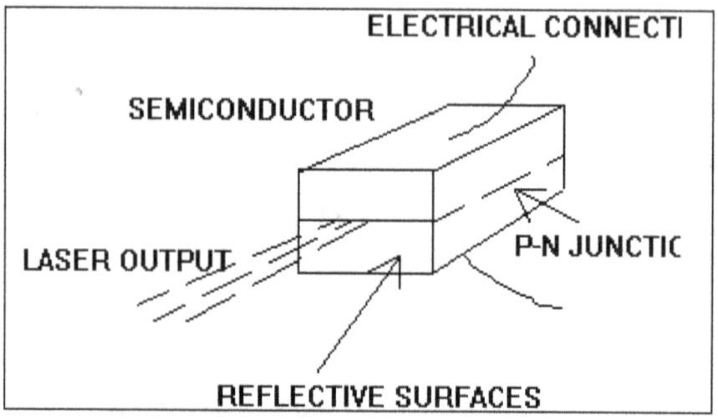

Semiconductor laser diagram.

Gas lasers consist of a gas filled tube placed in the laser cavity as shown in figure. A voltage (the external pump source) is applied to the tube to excite the atoms in the gas to a population inversion. The light emitted from this type of laser is normally continuous wave (CW). One should note that if brewster angle windows are attached to the gas discharge tube, some laser radiation may be reflected out the side of the laser cavity. Large gas lasers known as gas dynamic lasers use a combustion chamber and supersonic nozzle for population inversion.

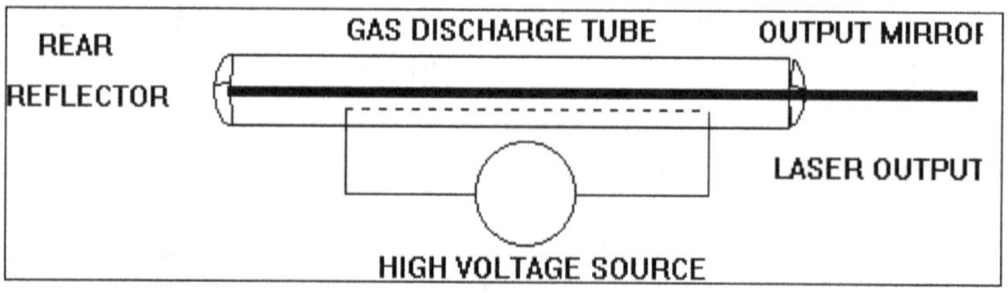

Gas laser diagram.

Figure shows a dye laser diagram. Dye lasers employ an active material in a liquid suspension. The dye cell contains the lasing medium. Many dyes or liquid suspensions are toxic.

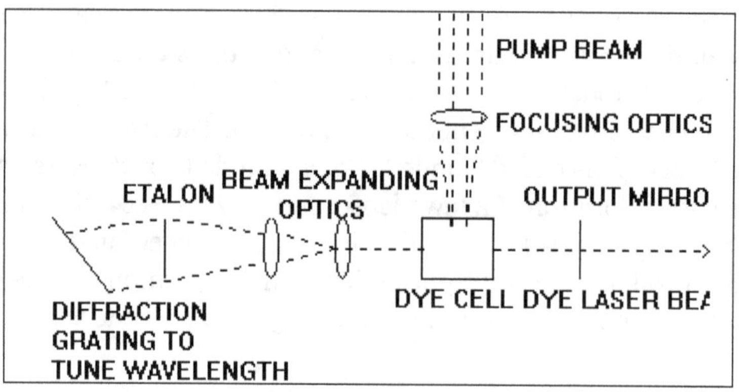

Common Dye Laser Diagram.

Free electron lasers such as in figure have the ability to generate wavelengths from the microwave to the X-ray region. They operate by having an electron beam in an optical cavity pass through a wiggler magnetic field. The change in direction exerted by the magnetic field on the electrons causes them to emit photons.

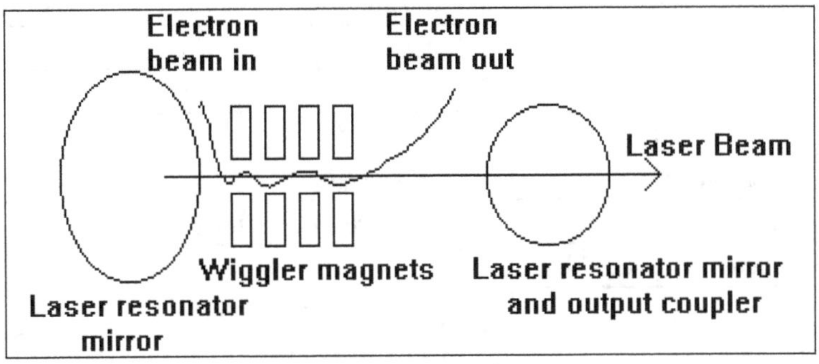

Free Electron Laser Diagram.

Laser beam geometries display transverse electromagnetic (TEM) wave patterns across the beam similar to microwaves in a wave guide. Figure shows some common TEM modes in a cross section of a laser beam.

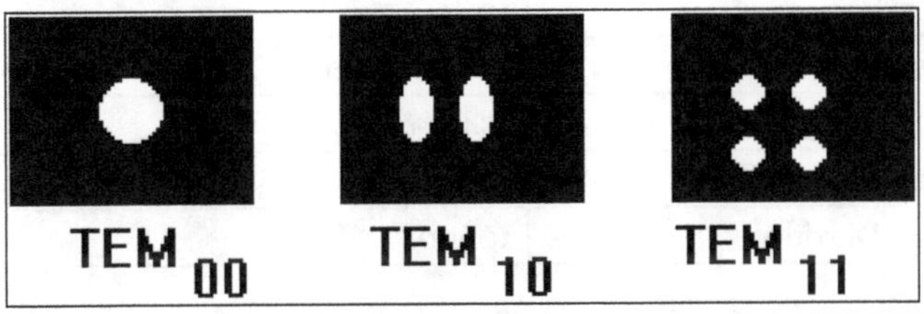

Common TEM laser beam modes.

A laser operating in the TEM_{10} mode could be considered as two lasers operating side by side. The ideal mode for most laser applications is the TEM_{00} mode and this mode is normally assumed to easily perform laser hazards analysis. Light from a conventional light source is extremely broadband (containing wavelengths across the electromagnetic spectrum). If one were to place a filter that would allow only a very narrow band of wavelengths in front of a white or broadband light source, only a single light color would be seen exiting the filter. Light from the laser is similar to the light seen from the filter. However, instead of a narrow band of wavelengths none of which is dominant as in the case of the filter, there is a much narrower linewidth about a dominant center frequency emitted from the laser. The color or wavelength of light being emitted depends on the type of lasing material being used. For example, if a Neodymium:Yttrium Aluminum Garnet (Nd:YAG) crystal is used as the lasing material, light with a wavelength of 1064 nm will be emitted. Table illustrates various types of material currently used for lasing and the wavelengths that are emitted by that type of laser. Note that certain materials and gases are capable of emitting more than one wavelength. The wavelength of the light emitted in this case is dependent on the optical configuration of the laser.

Table: Common lasers and their wavelengths.

LASER TYPE	WAVELENGTH (Nanometers)
Argon Fluoride	193
Xenon Chloride	308 and 459
Xenon Fluoride	353 and 459
Helium Cadmium	325 - 442
Rhodamine 6G	450 - 650
Copper Vapor	511 and 578
Argon	457 - 528 (514.5 and 488 most used)
Frequency doubled Nd:YAG	532
Helium Neon	543, 594, 612, and 632.8
Krypton	337.5 - 799.3 (647.1 - 676.4 most used)
Ruby	694.3
Laser Diodes	630 - 950
Ti:Sapphire	690 - 960
Alexandrite	720 - 780
Nd:YAG	1064
Hydgrogen Fluoride	2600 - 3000
Erbium:Glass	1540
Carbon Monoxide	5000 - 6000
Carbon Dioxide	10600

Light from a conventional light source diverges or spreads rapidly as illustrated in figure. The intensity may be large at the source, but it decreases rapidly as an observer moves away from the source.

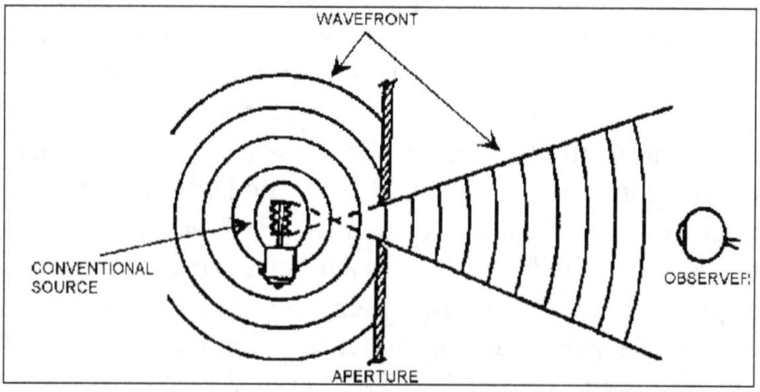

Divergence of Conventional Light Source.

In contrast, the output of a laser as shown in figure has a very small divergence and can maintain high beam intensities over long ranges. Thus, relatively low power lasers are able to project more energy at a single wavelength within a narrow beam than can be obtained from much more powerful conventional light sources.

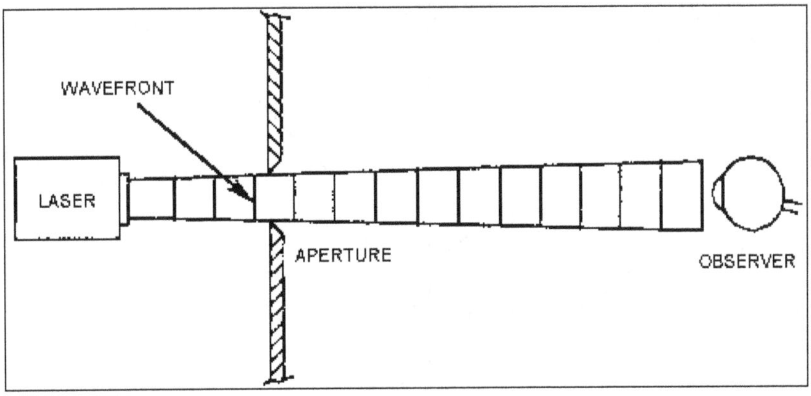

Divergence of Laser Source.

For example, a laser capable of delivering a 100 mJ pulse in 20 ns has a peak power of 5 million watts. A CW laser will usually have the light energy expressed in watts, and a pulsed laser will usually have its output expressed in joules. Since energy cannot be created or destroyed, the amount of energy available in a vacuum at the output of the laser will be the same amount of energy contained within the beam at some point downrange (with some loss in the atmosphere). Figure illustrates a typical laser beam. The amount of energy available within the sampling area will be considerably less than the amount of energy available within the beam. For example, a 100 mW laser output might have 40 mW measured within 1 cm^2 sample area. The irradiance in this example is 40 mW/ cm^2 .

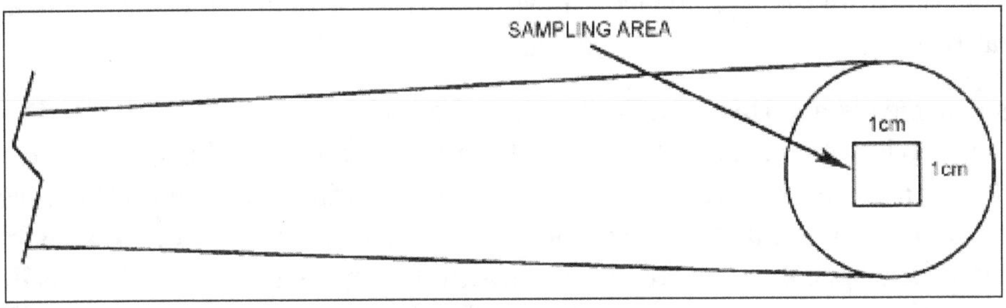

Illustration of Irradiance.

Characteristics of Materials

Materials can reflect, absorb, and/or transmit light rays. Reflection of light is best illustrated by a mirror. If light rays strike a mirror, almost all of the energy incident on the mirror will be reflected. figure illustrates how a plastic or glass surface will act on an incident light ray. The sum of energy transmitted, absorbed, and reflected will equal the amount of energy incident upon the surface.

A surface is specular (mirror-like) if the size of surface imperfections and variations are much smaller than the wavelength of incident optical radiation. When irregularities are randomly oriented and are much larger than the wavelength, then the surface is considered diffuse. In the intermediate region, it is sometimes necessary to regard the diffuse and specular components separately.

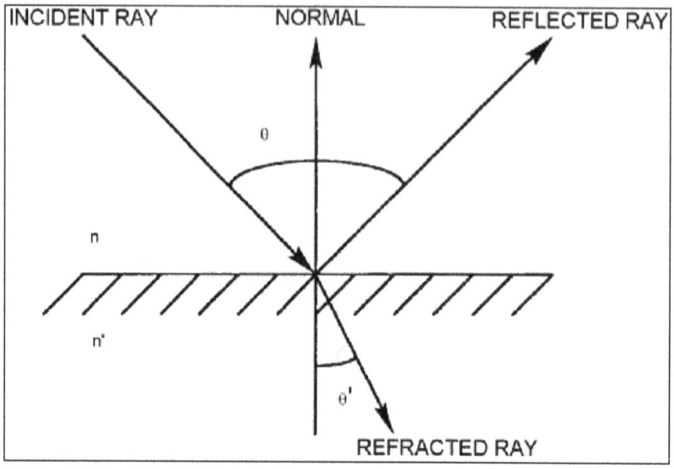

Light Ray Incident to Glass Surface.

A flat specular surface will not change the divergence of the incident light beam significantly. However, curved specular surfaces may change the divergence. The amount that the divergence is changed is dependent on the curvature of the surface. Figure demonstrates these two types of surfaces and how they will reflect an incident laser beam. The divergence and the curvature of the reflector have been exaggerated to better illustrate the effects. Note that the value of irradiance measured at a specific range from the reflector will be less after reflection from the curved surface than when reflected from the flat surface unless the curved reflector focuses the beam near or at that range.

A diffuse surface is a surface that will reflect the incident laser beam in all directions. The beam path is not maintained when the laser beam strikes a diffuse reflector. Whether a surface is a diffuse reflector or a specular reflector will depend upon the wavelength of the incident laser beam. A surface that would be a diffuse reflector for a visible laser beam might be a specular reflector for an infrared laser beam (e.g., CO_2). As illustrated in figure, the effect of various curvatures of diffuse reflectors makes little difference on the reflected beam.

If light is incident upon an interface separating two transmitting media (as an air-glass interface), some light will be transmitted while some will be reflected from the surface. If no energy is absorbed at the interface, T + R = 1 where T and R are the fractions of the incident beam intensity that are transmitted and reflected. T and R are called the transmission and reflection coefficients, respectively. These coefficients depend not only upon the properties of the material and the wavelength of the radiation, but also upon the angle of incidence. The amount of the incident light beam that is reflected and the amount transmitted through the material is further dependent on the polarization of the light beam.

The angle that an incident ray of radiation forms with the normal to the surface will determine the angle of refraction and the angle of reflection (the angle of reflection equals the angle of incidence). The relationship between the angle of incidence (ϕ) and the angle of refraction (ϕ) is:

$$n \cdot \sin(\phi) = n' \cdot \sin(\phi)$$

where n and n' are the indices of refraction of the media that the incident and transmitted rays move through, respectively.

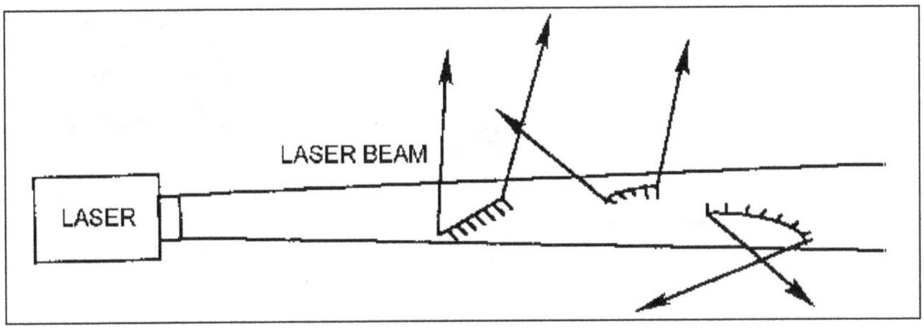

Specular Reflectors.

Principle of Working

In lasers, photons are interacted in three ways with the atoms:

- Absorption of radiation.
- Spontaneous emission.
- Stimulated emission.

Absorption of Radiation

Absorption of radiation is the process by which electrons in the ground state absorbs energy from photons to jump into the higher energy level.

The electrons orbiting very close to the nucleus are at the lower energy level or lower energy state whereas the electrons orbiting farther away from the nucleus are at the higher energy level. The electrons in the lower energy level need some extra energy to jump into the higher energy level. This extra energy is provided from various energy sources such as heat, electric field, or light.

Let us consider two energy levels (E_1 and E_2) of electrons. E_1 is the ground state or lower energy state of electrons and E_2 is the excited state or higher energy state of electrons. The electrons in the ground state are called lower energy electrons or ground state electrons whereas the electrons in the excited state are called higher energy electrons or excited electrons.

In general, the electrons in the lower energy state can't jump into the higher energy state. They need sufficient energy in order jump into the higher energy state.

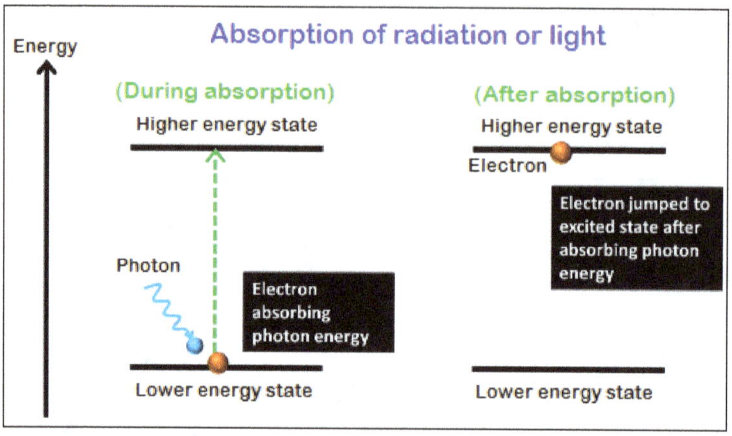

When photons or light energy equal to the energy difference of the two energy levels ($E_2 - E_1$) is incident on the atom, the ground state electrons gains sufficient energy and jumps from ground state (E_1) to the excited state (E_2).

The absorption of radiation or light occurs only if the energy of incident photon exactly matches the energy difference of the two energy levels ($E_2 - E_1$).

Spontaneous Emission

Spontaneous emission is the process by which electrons in the excited state return to the ground state by emitting photons.

The electrons in the excited state can stay only for a short period. The time up to which an excited electron can stay at higher energy state (E_2) is known as the lifetime of excited electrons. The lifetime of electrons in excited state is 10^{-8}second.

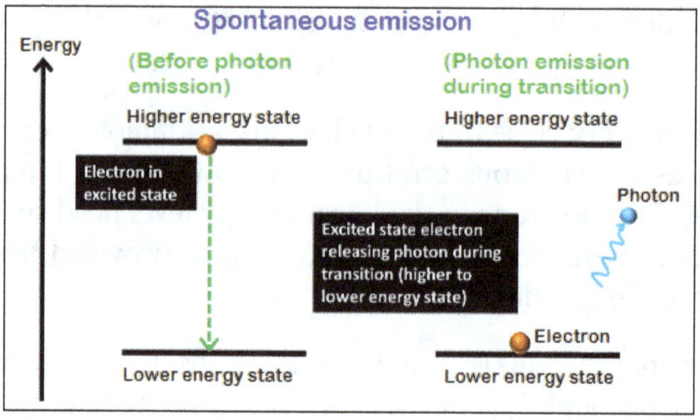

Thus, after the short lifetime of the excited electrons, they return to the lower energy state or ground state by releasing energy in the form of photons.

In spontaneous emission, the electrons move naturally or spontaneously from one state (higher energy state) to another state (lower energy state) so the emission of photons also occurs naturally. Therefore, we have no control over when an excited electron is going to lose energy in the form of light.

The photons emitted in spontaneous emission process constitute ordinary incoherent light. Incoherent light is a beam of photons with frequent and random changes of phase between them. In other words, the photons emitted in the spontaneous emission process do not flow exactly in the same direction of incident photons.

Stimulated Emission

Stimulated emission is the process by which incident photon interacts with the excited electron and forces it to return to the ground state.

In stimulated emission, the light energy is supplied directly to the excited electron instead of supplying light energy to the ground state electrons.

Unlike the spontaneous emission, the stimulated emission is not a natural process it is an artificial process.

In spontaneous emission, the electrons in the excited state will remain there until its lifetime is over. After completing their lifetime, they return to the ground state by releasing energy in the form of light.

However, in stimulated emission, the electrons in the excited state need not wait for completion of their lifetime. An alternative technique is used to forcefully return the excited electron to ground state before completion of their lifetime. This technique is known as the stimulated emission.

When incident photon interacts with the excited electron, it forces the excited electron to return to the ground state. This excited electron release energy in the form of light while falling to the ground state.

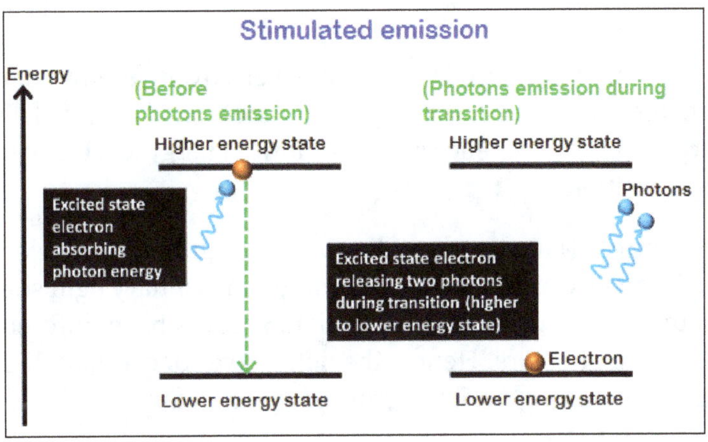

In stimulated emission, two photons are emitted (one additional photon is emitted), one is due to the incident photon and another one is due to the energy release of excited electron. Thus, two photons are emitted.

The stimulated emission process is very fast compared to the spontaneous emission process.

All the emitted photons in stimulated emission have the same energy, same frequency and are in phase. Therefore, all photons in the stimulated emission travel in the same direction.

The number of photons emitted in the stimulated emission depends on the number of electrons in the higher energy level or excited state and the incident light intensity.

It can be written as:

Number of emitted photons α Number of electrons in the excited state + incident light intensity.

Characteristics of a Laser

Laser light has four unique characteristics that differentiate it from ordinary light: these are:

- Coherence.
- Directionality.
- Monochromatic.
- High intensity.

Coherence

We know that visible light is emitted when excited electrons (electrons in higher energy level) jumped into the lower energy level (ground state). The process of electrons moving from higher energy level to lower energy level or lower energy level to higher energy level is called electron transition.

In ordinary light sources (lamp, sodium lamp and torch light), the electron transition occurs naturally. In other words, electron transition in ordinary light sources is random in time. The photons emitted from ordinary light sources have different energies, frequencies, wavelengths, or colors. Hence, the light waves of ordinary light sources have many wavelengths. Therefore, photons emitted by an ordinary light source are out of phase.

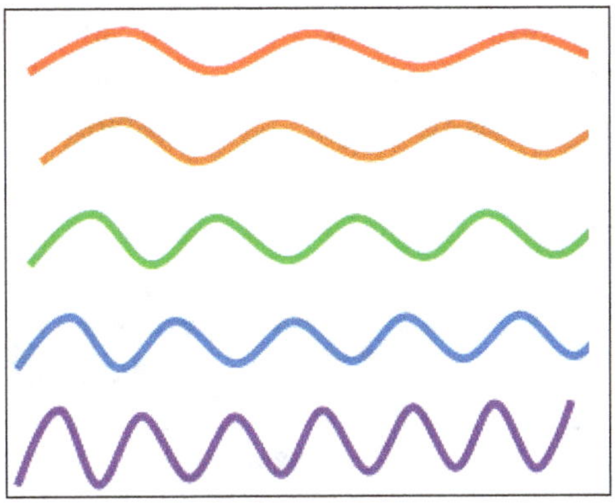

Incoherent light.

In laser, the electron transition occurs artificially. In other words, in laser, electron transition occurs in specific time. All the photons emitted in laser have the same energy, frequency, or wavelength. Hence, the light waves of laser light have single wavelength or color. Therefore, the wavelengths of the laser light are in phase in space and time. In laser, a technique called stimulated emission is used to produce light.

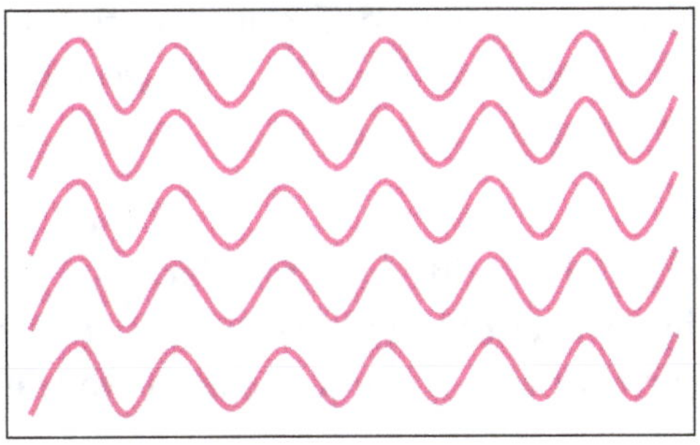

Coherent light waves.

Thus, light generated by laser is highly coherent. Because of this coherence, a large amount of power can be concentrated in a narrow space.

Directionality

In conventional light sources (lamp, sodium lamp and torchlight), photons will travel in random direction. Therefore, these light sources emit light in all directions.

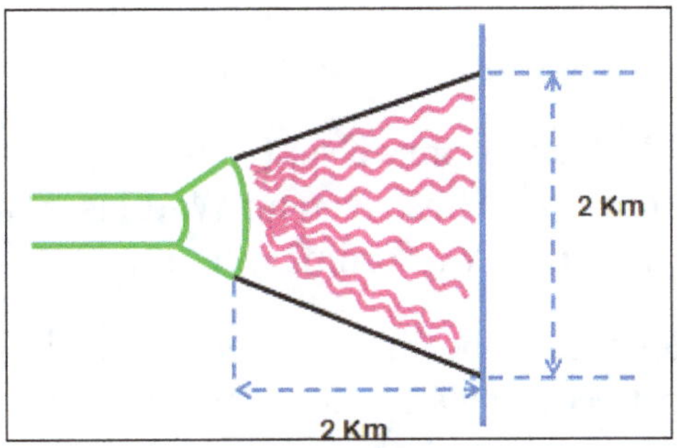

Ordinary light.

On the other hand, in laser, all photons will travel in same direction. Therefore, laser emits light only in one direction. This is called directionality of laser light. The width of a laser beam is extremely narrow. Hence, a laser beam can travel to long distances without spreading.

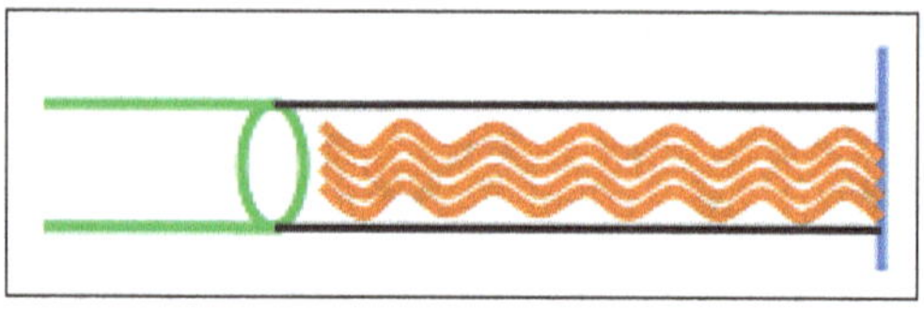

Laser light.

If an ordinary light travels a distance of 2 km, it spreads to about 2 km in diameter. On the other hand, if a laser light travels a distance of 2 km, it spreads to a diameter less than 2 cm.

Monochromatic

Monochromatic light means a light containing a single color or wavelength. The photons emitted from ordinary light sources have different energies, frequencies, wavelengths, or colors. Hence, the light waves of ordinary light sources have many wavelengths or colors. Therefore, ordinary light is a mixture of waves having different frequencies or wavelengths.

On the other hand, in laser, all the emitted photons have the same energy, frequency, or wavelength. Hence, the light waves of laser have single wavelength or color. Therefore, laser light covers a very narrow range of frequencies or wavelengths.

High Intensity

You know that the intensity of a wave is the energy per unit time flowing through a unit normal area. In an ordinary light source, the light spreads out uniformly in all directions.

If you look at a 100 Watt lamp filament from a distance of 30 cm, the power entering your eye is less than 1/1000 of a watt.

In laser, the light spreads in small region of space and in a small wavelength range. Hence, laser light has greater intensity when compared to the ordinary light.

If you look directly along the beam from a laser (caution: don't do it), then all the power in the laser would enter your eye. Thus, even a 1 Watt laser would appear many thousand times more intense than 100 Watt ordinary lamp.

Thus, these four properties of laser beam enable us to cut a huge block of steel by melting. They are also used for recording and reproducing large information on a compact disc (CD).

Types of Laser

Lasers can have a very broad range of emission powers and frequencies depending on the application they are required for, and different types of laser require different lasing mediums to function.

This means that there are thousands of different types of lasers which are divided into different types according to the lasing medium they use. The most common 'types' of laser are gas, liquid, solid-state and semiconductor lasers.

Gas Lasers

Gas lasers use an electric current discharged through a gas medium to produce light. The most common gas laser is the helium-neon laser, but others include argon ion lasers, carbon dioxide and carbon monoxide lasers, nitrogen lasers and hydrogen lasers. The gas that is used determines the wavelength that the laser produces. Gas lasers are used where a high beam quality and long coherence length is needed, and as high powers are possible, they are frequently used for the cutting of hard materials.

Excimer lasers are a type of gas laser that uses a combination of both reactive and inert gases as a lasing medium. For example, chlorine or fluorine can be mixed with gases such as argon, xenon or krypton. The name is short for 'excited dimer', because they produce a pseudo dimer molecule when stimulated in the ultraviolet range. Excimer lasers are frequently used for semiconductor photolithography and LASIK eye surgery.

Liquid Lasers

The most common type of liquid lasers are dye lasers, which use organic dyes as the laser medium. Dye lasers generate light from excited energy states of organic dyes which have been dissolved in liquid solvents. Dyes can be used at a much broader wavelength than other types of laser due to the variety of dyes available and how easily tuned they can be.

Dyes that are often used include stilbene, coumarin and rhodamine 6G, but many more are available. Liquid lasers are used mainly for research and in laser medicine.

Solid-state Lasers

Solid-state lasers use solids such as glass or crystaline materials as laser mediums, with ions of rare earth elements introduced as impurities via doping. Common materials include sapphire, neodymium-doped yttrium aluminum garnet, neodymium-doped glass and ytterbium-doped glass.

In solid-state lasers, light energy is used as a pumping source. Using solid state lasers is a common method when removing tattoos.

Semiconductor Lasers

The final type of lasers are semi-conductor lasers, which are also known as laser diodes. In semiconductor lasers, the junction of the semiconductor diode is the laser medium. Semiconductor lasers also use an electrical pumping source, which stimulates the emission of light.

Semiconductor lasers are the lasers used frequently in consumer goods as they are cheap, compact, and use little power.

Categorizing Lasers by their Pulse Duration

Lasers can also be characterized by the duration of the laser emission, which can be continuous wave, pulsed, Q-Switch, repetitively pulsed or mode locked.

Continuous lasers operate with a stable average beam power that can be adjusted in high power systems and is fixed in low power gas lasers.

Single pulsed lasers are also known as long pulse or normal mode lasers. They have pulse durations of a few hundred microseconds to a few milliseconds. Single-pulsed Q-Switch lasers use a Q-switch cell to allows the laser media to store a maximum of potential energy before emission occurs in single pulses.

Repetitively pulsed lasers, also known as scanning lasers, use pulsed lasers at fixed or variable pulse rates that range from a few pulses per second up to as 20,000 pulses per second.

Mode locked lasers use resonant modes of the optical cavity to affect the characteristics of the output beam. When the phases of different frequency modes are locked together, the different modes interfere with each another to generate a beat effect, resulting in a laser output with regularly spaced pulsations.

Categorizing Lasers by their Safety Rating

A final way of classifying lasers is splitting them into different categories based on their hazards. The power, pulse duration and wavelength combined determines the classification for the rules and regulations of safety for laser users.

The different classifications are Class 1, Class 1M, Class 2, Class 2M, Class 3R, Class 3M and Class 4. Lasers can be visible, invisible or both, so precautions need to be taken for the higher classifications of lasers.

The safest laser classification is Class 1, which is eye-safe under all operating conditions. Class 1M is safe for viewing directly with the naked eye but is considered to be hazardous if viewed with the aid of optical instruments. Radiation in classes 1 and 1M can be visible, invisible or both. Both classes 1 and 1M are in enclosures that limits access to the laser radiation, for example CD Rom players.

Class 2 lasers are visible lasers that are considered safe for accidental viewing, but they should not be looked at deliberately for longer than 0.25 seconds. Like Class 1M, Class 2M is considered to be more hazardous if viewed with the aid of optical instruments. An example of a class 2 laser is a supermarket scanner.

Class 3R lasers are considered low risk but potentially hazardous. The output is between 1 and 5 times Class 1, and they are commonly used in laser pointers. Class 3B is considered to be dangerous compared to previous categories. Viewing of the diffuse reflection is safe, but they are hazardous to the eye or skin if the direct beam is viewed. For a continuous wave laser, the maximum output into the eye must not exceed 500mW.

Class 4 is the most dangerous classification, as they are capable of setting fire to materials as well as causing eye and skin injuries through direct and reflected exposure.

Permissions

All chapters in this book are published with permission under the Creative Commons Attribution Share Alike License or equivalent. Every chapter published in this book has been scrutinized by our experts. Their significance has been extensively debated. The topics covered herein carry significant information for a comprehensive understanding. They may even be implemented as practical applications or may be referred to as a beginning point for further studies.

We would like to thank the editorial team for lending their expertise to make the book truly unique. They have played a crucial role in the development of this book. Without their invaluable contributions this book wouldn't have been possible. They have made vital efforts to compile up to date information on the varied aspects of this subject to make this book a valuable addition to the collection of many professionals and students.

This book was conceptualized with the vision of imparting up-to-date and integrated information in this field. To ensure the same, a matchless editorial board was set up. Every individual on the board went through rigorous rounds of assessment to prove their worth. After which they invested a large part of their time researching and compiling the most relevant data for our readers.

The editorial board has been involved in producing this book since its inception. They have spent rigorous hours researching and exploring the diverse topics which have resulted in the successful publishing of this book. They have passed on their knowledge of decades through this book. To expedite this challenging task, the publisher supported the team at every step. A small team of assistant editors was also appointed to further simplify the editing procedure and attain best results for the readers.

Apart from the editorial board, the designing team has also invested a significant amount of their time in understanding the subject and creating the most relevant covers. They scrutinized every image to scout for the most suitable representation of the subject and create an appropriate cover for the book.

The publishing team has been an ardent support to the editorial, designing and production team. Their endless efforts to recruit the best for this project, has resulted in the accomplishment of this book. They are a veteran in the field of academics and their pool of knowledge is as vast as their experience in printing. Their expertise and guidance has proved useful at every step. Their uncompromising quality standards have made this book an exceptional effort. Their encouragement from time to time has been an inspiration for everyone.

The publisher and the editorial board hope that this book will prove to be a valuable piece of knowledge for students, practitioners and scholars across the globe.

Index

www.ingramcontent.com/pod-product-compliance
Lightning Source LLC
LaVergne TN
LVHW080611210926
848485LV00003B/141
* 9 7 8 1 6 3 2 3 8 8 8 7 2 *